Contents

KU-759-806

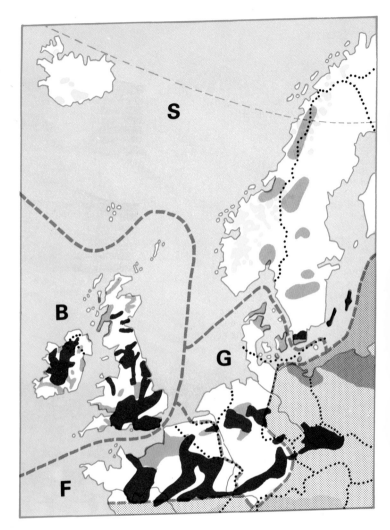

S

B

G

F

region not covered by this book

boundary between distribution areas

major outcrops of chalk and limestone

scattered outcrops of chalk and limestone

land over 1000m. (3250ft.)

The map does not show limestone in the Alps

Introduction

The Text

All the native grasses, sedges, rushes, ferns, horsetails and other fern allies of the north-western quadrant of Europe are described in the text, as well as a number of widely established aliens. Together with *The Wild Flowers of Britain and Northern Europe* (see Further Reading, p. 250), this book therefore comprises a complete field guide to the higher plants of Northern Europe. By higher plants is meant all vascular plants – plants whose tissue contains channels for conducting liquids. These can be subdivided into angiosperms (flowering plants), gymnosperms (conifers) – the two groups which comprise the spermatophytes or seed-plants – and pteridophytes, the ferns, clubmosses and horsetails.

The **area covered** by this book is the same as that of *The Wild Flowers of Britain and Northern Europe*, extending to Arctic Norway in the north and to the River Loire in western and central France and the River Danube in southern Germany in the south. Its southern boundary runs along the Loire from its mouth on the Atlantic coast of France eastwards to Dijon in eastern France and Basle in northern Switzerland and thence along the Danube and the foothills of the Alps to Munich, so as to include the Vosges and the Black Forest, but not the Jura and the Alps. At Munich it turns north across Germany through Regensburg, Bayreuth, Erfurt and Braunschweig, to reach the Baltic at Lübeck, and thence up the Gulf of Finland to the Arctic Ocean. Iceland is included. These boundaries are shown on the map opposite.

To indicate **distribution** in the text, the north-western segment of Europe has been divided into four regions. (1) Great Britain, Ireland and the Isle of Man; (2) France, Belgium, Luxembourg and the Channel Isles; (3) Germany, the Netherlands, and Denmark; and (4) Norway, Sweden, the Faeroes, and Iceland. See p. 7 for the symbols used to indicate these regions.

The maps (pp. 224–49), which should always be used in conjunction with the text, give distributions for the majority of plants in the guide. In fact, for the sake of completeness, they cover an area larger than the segment of north-western Europe described in the text – see the map opposite. For an explanation of the maps and the basis on which they have been compiled, see p. 223.

Description. Since the illustration of each plant is opposite its text, descriptive details are largely confined to points which cannot be easily illustrated, to facts which cannot be illustrated at all, and to stressing points which are crucial for identification. Each text description therefore includes certain standardised information, such as height, flower-

Introduction

ing time and habitat, and the major diagnostic features distinguishing the plant from other plants with which it might be confused are shown in italics. The text for each species is compiled as follows:

Group descriptions of families and genera (see p. 8) are given wherever possible, so as to avoid constant repetition of common characters, such as the solid unjointed stems of sedges and the long white hairs of wood-rushes. They should therefore always be read carefully and in conjunction with the species descriptions. The group descriptions refer only to species mentioned in the text.

English names mostly follow the list *English Names of Wild Flowers*, by J. G. Dony, C. M. Rob and F. H. Perring (1974), recommended by the Botanical Society of the British Isles. Where no English names exist we have in some cases translated those in August Garcke's *Illustrierte Flora* for Germany and neighbouring territories (1972).

Latin scientific names follow *Flora Europaea* (1964–1980), with a few later changes, mainly from the *Atlas of Ferns of the British Isles*, by A. C. Jermy, H. R. Arnold, Lynne Farrell and F. H. Perring (1978) and C. Page's *Ferns of Britain and Ireland* (1982).

British status: plants native to, or commonly naturalised in, Great Britain, Ireland and their associated islands are marked by an asterisk.

Number: each plant has a number corresponding with its number on the plate opposite, and may be referred to on the same page by this number. Where several plants are very similar, only one is illustrated, the others being described in the same paragraph. These **subsidiary species** are given the number of the main species and a letter, e.g. Snowy Wood-rush (2a) is subsidiary to White Wood-rush (2) on p. 182. The descriptions of all subsidiary species, however many are listed, refer back to the main species unless otherwise stated, and only the principal characters that differ are mentioned. On a few pages there has not been room to describe every plant that is illustrated, in which case the descriptions can be found below a rule on the previous or following page of text.

Hairiness or hairlessness, often an important clue to identification, is usually indicated.

Height is indicated as follows: *tall* means over 60 cm (2 ft); *medium*, 30–60 cm (1–2 ft); *short*, 10–30 cm (4–12 in); and *low*, 0–10 cm (0–4 in). Plant sizes can vary greatly, according to altitude, climate and soil. Plants are more likely to be found smaller than is indicated here than larger.

Status: annual or perennial status is shown for each species. Perennials are usually stouter than annuals and, of course, are much more likely to be seen above ground in winter. Only perennials spread vegetatively by rhizomes, stolons and similar structures.

Leaf-shapes (ferns): see Glossary (p. 220) for definition of terms such as pinnate.

Introduction

Inflorescence arrangements: see Glossary (p. 220) for definition of terms such as raceme.

Flowering time refers to the central part of the area, and may be earlier in the south, near the sea and in forward seasons, and later in the north, on mountains and in backward seasons.

Fruits are usually described only when important for identification.

Habitat: for definition of such terms as fen and bog, here used strictly, see Glossary (p. 220). The major mountain regions and areas with lime (chalk or limestone) in the soil are shown on the map on p. 4.

Distribution: the following symbols are used in the text to show whether a plant occurs, either commonly or uncommonly, within each region. If it has only a very few localities, 'rare' is added, and if it is mainly found, e.g. in the south of the region, the word 'southern' is also added.

T – Throughout the area covered by the book (see p. 5).
B – Great Britain, Ireland, Isle of Man.
F – France, Belgium, Luxembourg, Channel Isles.
G – Germany, the Netherlands, Denmark.
S – Norway, Sweden, Iceland, the Faeroes.

If parentheses enclose the symbol – for example, (G) – the plant is introduced, not native.

The Illustrations

Over 420 species are illustrated in colour, nearly all at life-size. Sometimes the distinctive character of a subsidiary species is also illustrated. The symbol □ is used after the scientific name of all subsidiary species with an illustration on the plate opposite.

Since the illustrations are arranged in scientific order, it will often be necessary to identify a plant *via* the keys. A plant should never be identified simply from the picture, especially since several similar species may be described under the one illustrated. Always remember, too, that a plant may sometimes be a different shade of colour from the one illustrated, and that some important identification features may not be of the type that can be shown in an illustration.

The illustrations in black and white beside the text show important diagnostic characters such as spikelets, fruits and spore-cases, many of which can only be satisfactorily studied with a hand lens. Here, they are illustrated many times larger than life-size; indeed, a few species can only be identified by such features. The symbol △ is used in the text after the name of the part of the plant that is illustrated.

Introduction

Classification and Scientific Nomenclature

The plants in this book are all either angiosperms (flowering plants) or pteridophytes (ferns, club-mosses, horsetails and their allies). The flowering plants are divided into two groups, monocotyledons and dicotyledons, whose seedlings have respectively one and two cotyledons (seed-leaves). Grasses, sedges and rushes are all monocotyledons, or 'monocots' in botanists' slang. Besides having only one seed-leaf, they all have mature leaves which are usually narrow and unstalked, often parallel-sided and nearly always parallel-veined; and flower parts almost always in multiples of three.

Like all other plants, monocotyledons and pteridophytes are classified into families, genera and species, groups of increasingly close affinity, in what, according to current scientific opinion, corresponds to the way in which they have evolved. Family names usually end in -aceae; grasses (Gramineae) are one of the few exceptions to this rule, though some American botanists call this family Poaceae. Individual scientific plant names are binomial; the name of the genus, or generic name, comes first, always with a capital initial, followed by the name of the species, or specific name, always with a small initial and, since it is an adjective, agreeing in gender with the generic name. This system, devised by the great eighteenth-century Swedish naturalist Linnaeus, is the international scientific nomenclature and, using Latin or Latinised Greek, enables naturalists in any country to know exactly what plant is being referred to. Thus British, French and German botanists all know Annual Meadow-grass, *Paturin annuel* and *Einjähriges Rispengras* as *Poa annua.*

Abbreviations used in the Text

Agg.	Aggregate
Incl.	Including
Sp.	Species (singular)
Spp.	Species (plural)
Ssp.	Subspecies
Var.	Variety
×	Indicates a hybrid, e.g. 1 × 2 is a hybrid between species 1 and 2.
*	Native or commonly naturalised in Britain.
□	This symbol after the name of a subsidiary species (see p. 6) indicates that it is illustrated in colour.
△a	This symbol after the name of a part of a plant (e.g. a spikelet or a fruit) indicates that it is illustrated in black and white.
B, F, G, S, and T	Found in Britain, France, Germany, Scandinavia, and Throughout respectively. For precise definition of these terms, see p. 5. Brackets indicate that the plant is introduced, not native.
WFBNE	*The Wild Flowers of Britain and Northern Europe* (see Further Reading, p. 250).

8

The Plants in This Book

Ferns (Pteridophyta: Filicopsida)

The ferns are part of the Division Pteridophyta of the plant kingdom,
though the term is used here more broadly to cover both the ferns
proper and their less conspicuous relatives such as the horsetails,
club-mosses and quillworts. The Pteridophytes are traditionally treated
as legitimate components of a *Flora* by botanists, who have almost
always excluded the mosses and liverworts (Bryophytes) and Algae,
although one group of Algae, the stoneworts or Charophytes, which
are water plants, used often to be found in *Floras*.

The Algae are entirely bound to water by their method of reproduction — they have spores that swim in water. The Bryophytes — mosses
and liverworts — also require moist conditions for fertilisation, but in
one part of their life cycle they produce dry spores that are blown about
by wind, and so they qualify as the most primitive true land-plants.
Significantly, however, the mosses have no proper system of transport-
ing water from roots (they do not have roots anyway) to leaves, an
essential for a really successful land-plant.

The Pteridophytes are the next group up the scale of evolved com-
plexity. They are well adapted to life on land in the main part of their life
cycle (fig. 1), but they retain an Achilles' heel. The spores produced on

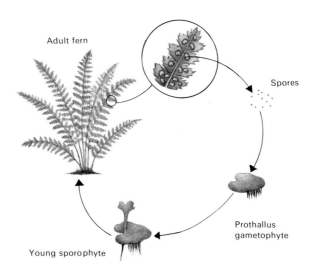

Adult fern

Spores

Prothallus
gametophyte

Young sporophyte

1. Fern life cycle

The Plants in this Book

the fronds of a fern germinate to give rise to a tiny, flimsy green disc called a prothallus; it looks a little like a small liverwort. This disc produces the gametes which, just as in the algae, must swim in a film of water to reach one another and bring about fertilisation. The new fern plant grows from this prothallus, but the gamete's need for water ties the adult fern to a habitat moist at least for long enough in the year to get through this phase.

Some ferns are remarkably successful in dry habitats, but they achieve this by vegetative growth (as anyone who has tried to control bracken will confirm) rather than by sexual reproduction. The great step forward made by the seed plants (Spermatophyta) was the invention of the seed and the abandonment of aquatic fertilisation. These higher plants still produce spores — tiny microspores (better known as pollen) which germinate, just like the spores of a fern, but on the stigma of a flower. There, protected from the environment, they can grow down as the pollen tube to fertilise the large megaspore (the ovule), and ultimately produce the seed. All this requires the pollen to reach the megaspore which is retained on the parent plant, and that necessitated the evolution of the flower.

Ferns then have no flowers, but they produce spores on the mature plant. Some, including all the true ferns, produce one type of spore. Others, such as quillworts, produce distinct large female spores (megaspores) and tiny male spores (microspores), the latter in much greater numbers, foreshadowing the seed plants.

Rushes and Wood-rushes (Juncaceae)

Rushes have the simplest flowers of the plants in this book. They may look obscure and 'difficult', but close examination (fig. 2) shows that they consist of two rings, each containing three 'tepals' or perianth segments (as there is no clear distinction between sepals and petals). The perianth is normally greenish or brown and not brightly coloured.

2. Rush flower

3. Inflorescence of *Juncus effusus*

Inside this perianth there are either one or two rings of three stamens, and in the centre of the flower there are three stigmas. As is characteristic of the monocotyledons everything is in multiples of three, whereas the dicotyledons go for fours and fives.

The flowers of rushes are in clusters or tight heads, either at the tip of a stem or apparently on the side of a stem (fig. 3). This is an illusion as the top part of the stem in these cases is really a bract, a leaf- or stem-like structure. Rushes *Juncus* have either flat hairless leaves or cylindrical stem-like leaves; wood-rushes *Luzula* always have flat leaves which bear long white hairs.

Rushes and wood-rushes are slow-growing plants typical of infertile soils, though some species are found on dry soils, others on wet soils. One or two rushes are annuals, often found in tracks and other wet places.

Sedges (Cyperaceae)

The structure of sedge flowers is less obvious to the untrained eye than that of rushes. They have no recognisable petals or perianth, though in some genera (e.g. cotton-grasses *Eriophorum*) the perianth consists of fine bristles that become very conspicuous in fruit. The basic structure is reduced to a small flap called a glume, enclosing a spikelet of flowers. In the largest genus *Carex*, and in *Kobresia*, the spikelets consist of a single male or female flower (fig. 4); in other words a male spikelet is simply a glume and two or three stamens, and a female spikelet is a glume, an ovary and two or three styles. In the other genera the spikelets contain both male and female flowers (fig. 5).

4. *Carex* sp. **5.** *Eriophorum* sp.

The spikelets are sometimes solitary on the ends of shoots, as in *Eleocharis* (fig. 6), *Scirpus cespitosus* and *Eriophorum vaginatum*, but more often they are gathered into heads or spikes. In *Carex* they are in spikes which either contain both male and female spikelets (fig. 7), or more often are all either wholly male or wholly female (fig. 8). In the

The Plants in this Book

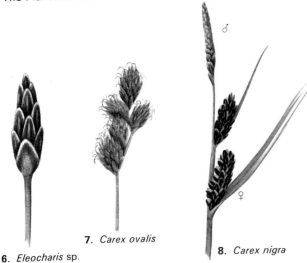

6. *Eleocharis* sp.

7. *Carex ovalis*

8. *Carex nigra*

latter case the spikes consisting of male spikelets only look quite different from the female spikes, particularly in fruit, and that provides an extremely important diagnostic feature. In fact it divides the genus into two subgenera — *Vignea* with all its spikes similar (male and female mixed up together), and *Carex* with the male spikes separate and usually on top of the flowerhead.

The fruits of sedges are often important for identification (fig. 9). In

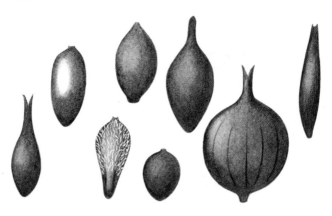

9. Sedges: assorted fruits

The Plants in this Book

Carex the fruit proper is enclosed in a capsule called the utricle, but for simplicity in this book the utricle is simply called the fruit. It often has a beak at its apex which may be notched. The colour and size of the glumes and colour of the fruit are also important characters.

Vegetative characteristics of sedges which are used for identification include the growth form — long creeping rhizomes or dense tussocks — and the width and sheath (around the stem) of the leaves.

Sedges are almost all perennials. They are typically slow-growing plants and are very good indicators of either infertile or waterlogged soils; one of the main floristic effects of agricultural improvement, involving drainage and fertilisers, has been the steep decline of many once widespread sedges.

Grasses (Gramineae)

Grasses, it has to be admitted, have extremely complex flowers, but once the underlying structure is understood they become much less alarming; they do need to be studied with a hand lens, however. Basically grass flowers, like those of sedges, are organised in spikelets (fig. 10), and the whole flowerhead consists of a number of spikelets. The spikelets may contain a single flower (in *Agrostis* bent-grasses for example) or two or many, or in some cases a mixture of fertile and sterile florets.

Each spikelet has at its base two scales (except that in some grasses one may be minute) and these more or less enfold the spikelet. Their sizes, particularly relative to each other, are important characters. These scales are called the glumes. Inside the glumes are the florets, each of which consists of two more scales, the lemma and the palea, which enclose the styles and stamens. Only the outer of these two scales, the lemma, is really important as a character. If there is just one floret in the spikelet, then there will be four scales, but if more then there will usually be an obvious central stalk or axis arising between the glumes and bearing the florets, each with its two scales, lemma and palea. Particularly important features to look for are: the presence of awns, long needle-like points arising from these scales, and (under a lens) whether the scales have nerves and how many.

These spikelets are borne in a number of different types of inflorescence. Sometimes they are simply arranged along the stalk in a spike, as in ryegrass *Lolium* (fig. 11), or a stalked spike (a raceme) as in false bromes *Brachypodium* (fig. 12), but mostly they are arranged in branching heads called panicles. Panicles can, however, look very different, from the loose spreading type of meadow-grasses *Poa* (fig. 13) or bents *Agrostis* (fig. 14) to the very dense head of foxtail *Alopecurus* (fig. 15) or the one-sided crested dogstail *Cynosurus* (fig. 16).

There are also plenty of good characters to look for on the vegetative parts of grasses; indeed grasses are in many ways easier to identify than other plants from vegetative parts alone, which is why a key to do this is given on p. 26. Growth form is, of course, important; grasses include everything from annuals (e.g. *Poa annua*) through deciduous perennials (*Molinia*) to almost woody perennials (bamboos). They

13

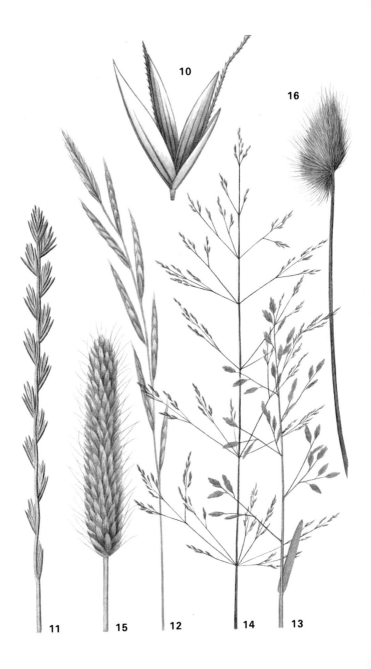

10

16

11   15   12   14   13

may form dense tufts or loose creeping mats, and the stems vary in hairiness, roundness and other characters — sometimes they are colourfully striped, as in *Holcus lanatus*.

The leaves provide one of the most important features — a tiny transparent flap between stem and leaf called the ligule (fig. 17). It is best seen by pulling the leaf slightly away from the stem. The ligule varies in size and shape and is sometimes replaced by hairs; all these make

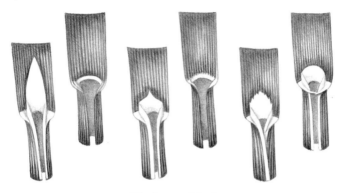

17. Assorted ligules

excellent diagnostic points. As you pull the leaf away from the stem, you will notice that it is attached to a sheath which enfolds the stem; this again is important. In particular you should note whether it forms a tube or has overlapping edges, and look to see whether there are little projections where sheath and leaf proper meet — these are called auricles.

The leaf blade is always long and narrow in grasses, compared for example to most dicotyledons, but still varies appreciably in size and shape, and also in being sometimes flat, sometimes rolled; sometimes hairy; sometimes ridged; and so on. The leaf tip is worth investigation; if you run your fingers to the end you can feel a distinct bump at the tip in some genera such as *Poa*; in others (e.g. *Agrostis*) the fingers run smoothly off it.

Grasses are ecologically quite unlike rushes and sedges. Indeed they are probably the most diverse group of plants from an ecological standpoint. They range from annual to perennials and are found in virtually all habitats, as a glance at the vegetative key (p. 26) will show. Unsurprisingly most grasses grow in grasslands, but grasses may be found on rocks, in woods, in water and so on. The success of grasses in grasslands rests on the fact that their leaves grow continuously from the base, so that as they are grazed off at one end (or cut or mown) they simply regenerate fresh tissue. This is why you can successfully mow your lawn but not your flowerbeds.

The Keys and how to use them

Since most of the species in this book (and virtually all of those that are at all common) are illustrated, many users will try to identify unknown plants by running quickly through the plates. This method is likely to work much of the time, at least in narrowing down your choice. It is not infallible, however, and some method of keying out plants is essential.

Many people are put off by the traditional dichotomous keys found in floras, though they are extremely efficient in the hands of a user familiar with the characters employed, and the professional botanist will always rely on them. They do have drawbacks, however; often they require knowledge which cannot be gleaned from your specimen (if it has no fruits for example) or ask questions which may baffle the uninitiated. This book tries a different approach by using a computer-generated key known as a *single-access key*. There are two single-access keys for the grasses, and one each for the sedges, rushes and ferns.

The Single-access Key

This is an entirely new sort of key. The principle here is the same as that by which computer keys work — you offer the computer a list of characters which you are sure that your specimen possesses, and it tells you the species which best fits that description. Species are thus identified by the possession of a unique set of characters, and not by one striking diagnostic feature. It must be emphasised, however, that no key can guarantee a correct identification. Having made a possible identification by using the keys, you must always return to the colour plates and the descriptions opposite them.

To work a single-access key you must produce a list of letters (known as a character string) which refers to a standard set of characters. These are given, with some explanatory notes, at the beginning of each key. All the characters can exist in two states, 1 and 2, and no intermediates are allowed. To identify an unknown plant you simply run down the list of characters and if the plant has character A in state 1 you score it as 'A'; if it is in state 2 you score it as 'a' but in fact omit it altogether from the string. You then repeat this with the remaining characters.

In the sedge key, for example, a sedge forming large raised tussocks (Ab), with leaves (c) 4 mm wide (D) and lower bract shorter than the flowerhead (e), with separate male and female flowers (f) in dissimilar spikes (g), each with two stigmas (h), the spikes clustered together (I), cylindrical (J), and unstalked (k), and the female glumes acute but not drawn out into a long point (I), dark brown (M), and the fruit clearly long-beaked (n) and not hairy (o) — would therefore score ADIJM.

The second part of each key consists of all possible combinations of these characters, listed alphabetically and giving the species possessing that particular combination. In the example, ADIJM refers uniquely to *Carex elata*. In some cases two or rarely more species possess the same character set, and then a small subsidiary key is included. This is indicated by the use of black dots (•). A single dot means that the character set has two alternatives. Sometimes one, or both, alternatives are further subdivided, and these entries are indicated by the use of two dots — and so on up to a maximum of four dots. In each case, start at the entry with a single dot and work down through the list of alternatives until the correct species is keyed out. Open arrowheads (▷) guide you from choice to choice.

Clearly in many cases a species may be variable for one of the key characters and either state may be possible. This is allowed for in the key by including two (or more) entries in the species list: if two characters vary in this way four entries occur, if three characters vary then eight, and so on.

The advantages of this method are

1 You only need to learn to work with a restricted group of characters. Obviously to complete your identification you will need to look at others, but those in the key are all useful diagnostic features in that group.

2 If a particular character cannot be determined on your specimen, you can simply look up both alternatives. Suppose that the sedge in the example above had no flowers so that you could not determine the number of stigmas (character H): simply look up both ADHIJM and ADIJM. The former is not listed, so that the latter must be correct. Of course this method will sometimes give a wrong possibility, and this must be checked against the descriptions opposite the plates.

3 If you have a specimen that is not in flower, you can still narrow down the list of possibles by using the first half of the character list only, since leaf and stem characters appear first. To improve the efficiency of this, these vegetative codes are split into habitat groups in the Vegetative Key to Grasses (p. 26).

Where to Start?

The single-access keys on pp. 20–43 are split into five — one each for the sedges, rushes and pinnate ferns and two for the grasses. Normally you will know exactly which one to go to, but in some cases there may be problems. The starter key below is intended to help you solve them. Three overall growth forms appear in this book:

1 Plants with finely divided, feathery (pinnate) leaves, often arising directly from the root-stock:

These are **ferns**, but beware a number of flowering plants such as Umbellifers (Carrot Family): see *WFBNE*.

Refer to the KEY TO PINNATE FERNS (p. 41).

The Keys and how to use them

2 Plants with long narrow grass-like leaves, either borne on a stem or in a tussock, sometimes of great size. These may be:

Grasses, which have rounded stems, usually rather soft leaves with a distinct, easily removed sheath around the stem, and the unique flowers described on p. 14. They are found in almost all habitats and may be annual or perennial.

See the KEYS TO GRASSES (pp. 20, 26).

Sedges, which typically have triangular stems (a few are rounded), channelled leaves which only rarely form a loose sheath around the stem, and flowers in spikelets as described on p. 13. They typically grow on dry or very wet soils, but not on the more fertile; almost all are perennials.

See the KEY TO SEDGES (p. 32).

Rushes, which have rounded, often hollow and pithy stems and rather similar leaves. They have the most 'flower-like' flowers (see p. 12), rather like a miniature brown lily. Otherwise more like sedges than grasses, but wood-rushes *Luzula* have flat grass-like leaves with long white hairs.

See the KEY TO RUSHES (p. 38).

3 Plants with green stems but no leaves (or only minute and scale-like leaves). The stems may bear no flowers, a cluster on the side, or a spike of flowers or a cone full of spores at the tip. These may be:

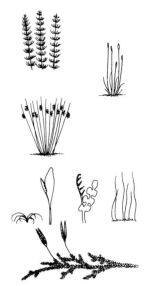

Horsetails, which often (but not always) have a cone at the tip, and have rings of scales up the stem.

See pp. 188–190.

Sedges, or strictly allies of the sedges such as club-rushes, often with triangular stems, and with a spike of flowers at the top. Usually in wet places.

See the KEY TO SEDGES (p. 32).

Rushes, which have rounded or ridged stems, and a cluster of flowers apparently on the side of the stem.

See the KEY TO RUSHES (p. 38).

A few unusual **ferns** and allies.

See FERN PLATES pp. 188–217.

Plants covered with fine scale-leaves: **clubmosses**.

See FERN PLATES, pp. 184–186.

18

Key to Grasses (Gramineae)

State 1	State 2
A Annual; often small plants with no non-flowering shoots, rarely found in closed vegetation	**a** perennial; often with non-flowering shoots, stolons, and rhizomes etc.
B Creeping, rhizomatous, stoloniferous or very loosely tufted or solitary	**b** clearly or densely tufted
C Ligule absent, a fringe of hairs, or less than 0.5 mm long	**c** ligule at least 0.5 mm long
D Ligule pointed	**d** ligule blunt
E Ligule toothed or ragged	**e** ligule entire
F Leaves less than 4 mm wide	**f** some leaves at least 4 mm wide
G Leaves or leaf-sheaths hairy	**g** no hairs

Key to Grasses (Gramineae)

H   Leaf-sheath with auricles

h   no auricles

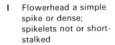

I   Flowerhead a simple spike or dense; spikelets not or short-stalked

i   flowers in a loose panicle

J   Spikelets 1-flowered

j   spikelets 2-flowered or more

K   Spikelets less than 5 mm long

k   spikelets more than 5 mm long

L   Awn more than 1 mm long, or bristles

l   awn less than 1 mm or absent

This key does not include bamboos (see p. 44). Several grasses have viviparous forms, where the flowers are replaced by tiny plantlets, some habitually. The most frequent include species of *Poa, Festuca,* and *Deschampsia,* particularly in upland areas. But *Cynosurus, Dactylis* and other lowland genera may have viviparous flowerheads, especially at the end of the season.

—	Poa chaixii
A	Briza maxima 66
ABEFGIL	Bromus lepidus 76
ABEFGL	Bromus spp. (see AEFGL)
ABFJK	Coleanthus subtilis 102
ABFJKL	Coleanthus subtilis 102
ABFIJKL	Alopecurus aequalis 100
ABGJK	Digitaria sanguinalis 110
ABJK	Coleanthus subtilis 102

ABJKL	Coleanthus subtilis 102
ACEF	Eragrostis cilianensis 106
ACEFG	Eragrostis minor 106
ACEFGIK	Crypsis alopecuroides 108
ACEFGK	Eragrostis pilosa 106
ACEFIK	Crypsis alopecuroides 108
ACEGIK	Crypsis aculeata 108
ACEIJKL	Setaria spp. 108
ACEIK	Crypsis aculeata 108

Key to Grasses (Gramineae)

ACFG   Micropyrum tenellum 52
ACFGIL   •upper leaf-sheath inflated: Vulpia fasciculata (incl. V. membranacea) 52
       •upper leaf-sheath not inflated ▷
       ••stem smooth: Vulpia ciliata 52
       ••stem ridged: Vulpia bromoides 52
ACFGL   •lower glume less than $\frac{3}{4}$ as long as upper: Vulpia bromoides 52
       •lower glume at least $\frac{3}{4}$ as long as upper: Micropyrum tenellum 52
ACFIJ   •plant less than 20 cm, spikes curved: Parapholis incurva 102
       •plant at least 20 cm, spikes straight: Parapholis strigosa 102
ACGL   Gaudinia fragilis 80
ACIK   Echinochloa colonum 112
ACKL   Echinochloa crus-galli 112
AD   Puccinellia rupestris 60
ADEFGJKL   Apera interrupta 64
ADEFGJL   Bromus sterilis 70
ADEFGL   Bromus sterilis 70
ADEFIL   Anthoxanthum aristatum 90
ADEGL   Bromus sterilis 70
ADEIJ   Phalaris canariensis 102
ADEIJK   Phalaris minor 102
ADEJKL   Apera spica-venti 64
ADF   •leaves wrinkled; inland: Poa annua 54
       •leaves smooth; saltmarshes: Puccinellia rupestris 60
ADFIJK   Polypogon maritimus 94
ADFIJKL   Gastridium ventricosum 94
ADFIKL   Aira praecox 88
ADFKL   Aira caryophyllea 88
ADFJK   •spikelets rounded; north to Holland: Milium vernale 104
       •spikelets flattened; Scandinavia: Agrostis clavata 92
ADFK   Poa annua 54
ADGL   Ventenata dubia 84

ADIJKL   Polypogon monspeliensis 94
ADIJL   Cynosurus echinatus 62
ADIL   Cynosurus echinatus 62
ADK   Briza minor 66
AEFGIL   Bromus lepidus 76
AEFGKL   Vulpia unilateralis 52
AEFGL   •spikelets broadest at top, awn long ▷
       ••some branches with 4 spikelets: Bromus tectorum 70
       ••branches with no more than 2 spikelets: Bromus madritensis 70
       •spikelets narrower at top, awn short ▷
       ••flowerhead dense: Bromus hordeaceus 76
       ••flowerhead loose, up to 30 cm long: Bromus arvensis (incl. B. squarrosus, B. japonicus) 74
       ••flowerhead loose, up to 15 cm long: Bromus racemosus 74
AEFI   Desmazeria rigida 54
AEGL   •spikelets hidden by glumes ▷
       ••spikelets drooping: Avena sterilis 84
       ••spikelets erect: Ventenata dubia 84
       •spikelets longer than glumes ▷
       ••spikelets widest at top ▷
       •••flowerhead loose, spreading: Bromus diandrus 70
       •••flowerhead denser, erect: Bromus rigidus 70
       ••spikelets narrower at top ▷
       •••spikelets hairy: Bromus hordeaceus 76
       •••spikelets hairless ▷
       ••••flowerhead upright, narrow: Bromus racemosus 74
       ••••flowerhead nodding, broad: Bromus commutatus 74

Key to Grasses (Gramineae)

AEHI	•spikelets more than 11 mm: Lolium temulentum 50
	•spikelets 8–11 mm: Lolium remotum 50
AEHIL	Lolium temulentum 50
AEI	Desmazeria marina 54
AEIJ	Zea mays 110
AEIJL	Alopecurus myosuroides 100
AEL	Bromus secalinus 74
AF	Puccinellia rupestris 60
AFGHIJL	Hordeum murinum 82
AFGIL	Vulpia myuros 52
AFGL	Vulpia myuros 52
AFIJK	•flowerhead dense, cylindrical: Phleum arenarium 98
	•spike simple, with spaced spikelets: Mibora minima 64
AFIJKL	Alopecurus rendlei 100
AFIJL	Hordeum marinum 82
AFJKL	Coleanthus subtilis 102
AFK	Poa infirma 54
AFL	Avena brevis 84
AGHI	Triticum aestivum 80
AGHIJL	Hordeum murinum 82
AGHIL	Triticum spp. 80
AGIJL	Lagurus ovatus 86
AGJK	Panicum miliaceum 110
AGL	•branches of flowerhead one-sided: Avena strigosa 84
	•flowerhead spreading▷
	••spikelets with brown hairs: Avena fatua 84
	••spikelets without hairs: Avena sativa 84
AHI	Lolium multiflorum 50
AHIJL	•spikelets in 2 rows: Hordeum distichon 82
	•spikelets in 4–6 rows: Hordeum vulgare 82
AHIL	•awn short; fodder grass: Lolium multiflorum 50
	•awn long; cereal crop: Secale cereale 80
AHL	Lolium multiflorum 50
AI	Lolium rigidum 50
AIJK	Digitaria ischaemum 110
AJK	Digitaria ischaemum 110
AL	•spikelets 10–12 mm: Avena brevis 84

	•spikelets 20–30 mm: Avena nuda 84
B	•in or by water▷
	••spikelets with many florets, widespread: Glyceria plicata 68
	••spikelets with 3–4 florets; NE Europe: Scolochloa festucacea 68
	•damp woods; NE Europe: Poa remota 56
BC	Melica nutans 66
BCEFJK	Cynodon dactylon 110
BCEI	Phragmites australis 102
BCEJ	•robust, ligule hairs long: Spartina anglica 112
	•slighter, ligule hairs short: Spartina maritima 112
BCF	Melica nutans 66
BCFGI	Elymus farctus 78
BCFGL	Festuca diffusa 48
BCFI	Elymus pycnanthus 78
BCFL	Festuca rubra (incl. F. richardsonii) 48
BCGHI	Elymus hispidus 78
BCGI	Elymus farctus 78
BCGJ	Melica uniflora 66
BCHI	Elymus repens (incl. E. pungens) 78
BCI	Elymus pycnanthus 78
BD	•2 anthers; NE Europe: Glyceria lithuanica 68
	•3 anthers; common: Glyceria maxima 68
BDEJK	Catabrosa aquatica 64
BDEK	Catabrosa aquatica 64
BDGIJ	Ammophila arenaria 94
BDF	•florets dark purple, often viviparous: Poa arctica 58
	•florets green, never viviparous: Poa supina 54
BDFJK	Agrostis stolonifera 92
BDFJKL	Agrostis canina 92
BDFK	•ligule at least 3 mm: Poa trivialis 56
	•ligule less than 3mm: Poa arctica 58
BDGIJL	× Ammocalamagrostis baltica 94
BDK	Poa trivialis 56
BE	Bromus inermis 7˄

Key to Grasses (Gramineae)

BEF	Arctophila fulva 62
BEFK	Arctophila fulva 62
BEGJL	•stems with 4 or more nodes; widespread: Calamagrostis canescens 96
	•stem with 2 nodes; Scandinavia: Calamagrostis lapponica 96
BEIJ	Phalaris arundinacea 102
BEIL	Brachypodium pinnatum 72
BEIJKL	×Agropogon littoralis 94
BEJ	Phalaris arundinacea 102
BEJK	Agrostis gigantea 92
BEJKL	Calamagrostis villosa 96
BEJL	•stems unbranched; mountains: Calamagrostis villosa 96
	•stems branched; wetlands: Calamagrostis purpurea 96
BF	•spikes nodding, solitary on long stalks: Briza media 66
	•not so▷
	••saltmarshes▷
	•••leaves 30 mm long, Arctic: Puccinellia phryganodes 60
	•••leaves 30 mm long, common: Puccinellia maritima 60
	••inland▷
	•••stems flattened: Poa compressa 54
	•••stems round: Poa pratensis (incl. P. angustifolia, P. alpigena and P. subcaerulea) 56
	••alpine; spikelets viviparous: Poa alpina 58
BFGI	Koeleria pyramidata 86
BFGIK	Koeleria macrantha 86
BFI	Elymus pycnanthus 78
BFIL	Brachypodium pinnatum 72
BFIJK	Polypogon viridis 94
BFJK	Agrostis capillaris 92
BFK	•spikelets nodding, solitary, on long stalks: Briza media 66

	•not so▷
	••stems flattened: Poa compressa 54
	••stems round: Poa pratensis (incl. P. angustifolia, P. alpigena and P. subcaerulea) 56
BFL	Festuca juncifolia 48
BGJK	Leersia oryzoides 112
BGKL	Holcus mollis 90
BGL	Holcus mollis 90
BHI	Leymus arenarius 78
BIJK	Polypogon viridis 94
BIJKL	Alopecurus geniculatus 100
BIL	Brachypodium pinnatum 72
BJK	Polypogon viridis 94
BKL	Sorghum halepense 112
BL	Sorghum halepense 112
CEFG	Danthonia decumbens 106
CEG	Molinia caerulea 106
CEGJ	Molinia caerulea 106
CEIJ	Calamagrostis varia 96
CEIJK	Calamagrostis varia 96
CEJ	Calamagrostis varia 96
CEJK	Calamagrostis varia 96
CF	•lemma shortly awned: Festuca amethystina 58
	•lemma awnless▷
	••spikelets few, large: Melica picta 66
	••spikelets many, small: Poa nemoralis 58
CFG	Danthonia decumbens 106
CFGIK	Koeleria vallesiana 86
CFHL	•spikelets viviparous: Festuca vivipara 48
	•spikelets normal▷
	••leaves green, common: Festuca ovina (incl. F. tenuifolia) 48
	••leaves bluish, rare: Festuca caesia 48
CFI	Sesleria albicans 66
CFJK	Poa nemoralis 58
CFJL	Achnatherum calamagrostis 104
CFL	•sheaths fibrous: Festuca heterophylla 48
	•sheaths not fibrous: Festuca nigrescens 48
CI	Sesleria albicans 66
CIL	Sesleria caerulea 66

Key to Grasses (Gramineae)

D •coastal muds and sands:
Puccinellia rupestris 60
•freshwater▷
••leaf-sheaths smooth,
lemma untoothed:
Glyceria fluitans 68
••leaf-sheaths rough,
lemma toothed:
Glyceria declinata 68

DEFJKL Agrostis vineale 92
DEHL Bromus benekenii 72
DF •coastal muds and sands:
Puccinellia rupestris 60
•freshwater: Glyceria
declinata 68
•alpine rocks; often
viviparous▷
••leaves more than 2 mm
wide: Poa alpina 58
••leaves less than 2 mm
wide: Poa ×
jemtlandica 58

DFJKL Agrostis curtisii 92
DFK Poa bulbosa 54
DFKL •some florets male only:
Hierochloë alpina 90
•all florets
hermaphrodite▷
••spikelets glossy▷
•••flowerhead to 15 cm
across: Deschampsia
media 88
•••flowerhead to 10 cm
across; rare:
Deschampsia setacea
88
•••flowerhead to 8 cm
across; common:
Deschampsia flexuosa
88
••spikelets dull:
Corynephorus
canescens 90

DFL Deschampsia flexuosa 88
DFGL Avenula pubescens 84
DFIJK Phippsia algida 60
DFIJKL Phleum pratense ssp.
bertolonii 98
DFJK Phippsia concinna 60
DFK •florets dark purple: Poa
arctica 58
•florets green: Poa laxa
58
DFKL Deschampsia cespitosa
88
DFL Avenula pratensis 84

DGL Avenula pubescens 84
DJK Milium effusum 104
DJL Calamagrostis
pseudophragmites 96
DK Hierochloë odorata 90
DKL Hierochloë australis 90
DL Arrhenatherum elatius 84

E Festuca altissima 46
EF Poa badensis 58
EFGIL Anthoxanthum odoratum
90
EFGJL Stipa pulcherrima 104
EFGL Trisetum flavescens 86
EFJ Melica ciliata 64
EFL Bromus erectus 72
EGIL Anthoxanthum odoratum
90
EGJKL •leaves more than 12 mm
broad: Cinna latifolia
64
•leaves less than 12 mm
broad: Calamagrostis
chalybaea 96
EHL Bromus ramosus 72
EIJL Calamagrostis epigeios 96
EIL •large plant, stems
flattened at base:
Dactylis glomerata 62
•smaller, stems round:
Anthoxanthum
odoratum 90
EJ Melica transsilvanica 66
EJL Calamagrostis
pseudophragmites 96
EL Bromus carinatus 74

F •alpine plants▷
••viviparous: Poa arctica
58
••not viviparous: Poa
glauca 58
•saltmarshes▷
••lemma less than 2.5
mm▷
•••spikelets at tips of
branches: Puccinellia
distans 60
•••spikelets all along
branches: Puccinellia
fasciculata 60
••lemma at least 2.5 mm:
Puccinellia rupestris 60
FGHIJKL Hordeum jubatum 82
FGHIJL Hordeum secalinum 82
FGIK Koeleria glauca 86

Key to Grasses (Gramineae)

FGIL Trisetum spicatum 86
FGJKL Calamagrostis stricta 96
FGJL •wet places; Scotland:
  Calamagrostis scotica
  96
  •dry steppe; C. Europe:
  Stipa spp. 104
FGL Trisetum subalpestre 86
FIJK •dry soils; southern:
  Phleum phleoides 98
  •mountain grassland,
  flushes: Alopecurus
  alpinus 100
FIJKL •stems swollen at base:
  Alopecurus bulbosus
  100
  •stems not swollen:
  Polypogon viridis 94
FIK Cynosurus cristatus 62
FIKL Cynosurus cristatus 62
FJKL Agrostis mertensii 92
FJL Nardus stricta 106
FK •alpine▷
  ••leaves less than 2 mm
  wide: Poa flexuosa 58
  ••leaves at least 2 mm
  wide: Poa glauca 58
  •marshes and freshwater:
  Poa palustris 56
  •saltmarshes▷
  ••spikelets at tips of
  branches: Puccinellia
  distans 60
  ••spikelets all along
  branches: Puccinellia
  fasciculata 60
FKL Vahlodea atropurpurea
  88

GH Bromus wildenowii 74
GHIJL Hordelymus europaeus 80
GIL •leaves yellowish, softly
  hairy: Brachypodium
  sylvaticum 72
  •leaves green, roughly
  hairy: Elymus caninus
  (incl. E. mutabilis, E.
  alaskanus) 78
GJL Calamagrostis
  arundinacea 96

GKL Holcus lanatus 90
GL Holcus lanatus 90

H •some branches of
  flowerhead with 1–2
  spikelets: Festuca
  pratensis 46
  •all branches with at
  least 2 spikelets:
  Festuca arundinacea 46
HI •no lower glume:
  Lolium perenne 50
  •some spikelets with 2
  glumes: ×Festulolium
  loliaceum 50
HL Festuca gigantea 46
HIL •no lower glume: Lolium
  multiflorum 50
  •some spikelets with 2
  glumes: ×Festulolium
  braunii 50
HL Bromus ramosus 72

IJK Alopecurus alpinus 100
IJKL •awn on glumes▷
  ••flowerhead rounded;
  alpine: Phleum alpinum
  98
  ••flowerhead cylindrical;
  lowland: Phleum
  pratense 98
  •awn on lemma:
  Alopecurus pratensis
  100
IJL Alopecurus pratensis 100
IL Elymus caninus (incl. E.
  mutabilis, E. alaskanus)
  78

J Calamagrostis varia 96
JK Milium effusum 104
JKL Dichanthium ischaemum
  110

L •spikelets 25–30 mm:
  Bromus carinatus 74
  •spikelets 6–12 mm:
  Arrhenatherum elatius
  84

Vegetative Key to Grasses (Gramineae)

This key is arranged by habitat, since anyone wishing to identify grasses on vegetative characters only is likely to be working in the field. This subdivision by habitat enables the codes in the main key (p. 20) that refer to vegetative characters (letters A to H) to be used without the floral characters (letters I to L), without producing unmanageably large groups. Where several species do share the same code, no subsidiary keys are given here, as in most cases a quick glance at the text and plate will make the final identification clear. To assist that search, however, rare species and non-British species are listed in italic type. (S) and (N) after the names of non-British species refer to southern and northern distributions in Europe.

Mobile dunes, shingle banks, sandy shores

ACFG	*Vulpia ciliata* 52
AE	*Desmazeria marina* 54
B	Elymus pycnanthus 78
BC	Elymus pycnanthus 78
BCF	Elymus pycnanthus 78
	Festuca rubra 48
BCFG	Elymus farctus 78
BCG	Elymus farctus 78
BDE	Catabrosa aquatica 64
BDG	Ammophila arenaria 94
BF	Elymus pycnanthus 78
	Festuca juncifolia 48
BH	Leymus arenarius 78
DF	*Poa bulbosa* 54
	Corynephorus canescens 90

Fixed dunes

ACFG	*Vulpia fasciculata* 52
	Vulpia membranacea (S) 52
ADF	Aira praecox 88
AF	*Mibora minima* 64
	Hordeum marinum 82
	Phleum arenarium 98
AG	*Lagurus ovatus* (S) 86
BCF	Festuca rubra 48
BDF	Agrostis stolonifera 92
BDG	×*Ammocalamagrostis baltica* 94
BF	Briza media 66
DF	Poa bulbosa 54
	Corynephorus canescens 90

Bare coastal mud

AD	*Puccinellia rupestris* 66
ADF	*Puccinellia rupestris* 60
AF	*Puccinellia rupestris* 60
BCE	Spartina spp. (especially S. anglica) 112
BE	×*Agropogon littoralis* 92
BF	*Festuca juncifolia* 48
	Puccinellia maritima 60
D	*Puccinellia rupestris* 60
DF	*Puccinellia rupestris* 60
F	Puccinellia distans 60
	Puccinellia fasciculata 60
	Puccinellia rupestris 60

Saltmarshes

ACF	Parapholis strigosa 102
	Parapholis incurva 102
AD	*Polypogon monspeliensis*
	Puccinellia rupestris 60
ADF	Polypogon maritimus 94
	Puccinellia rupestris 60
AF	*Puccinellia rupestris* 60
	Crypsis aculeatus (S) 108
BCE	Spartina maritima 112
	Spartina anglica 112
BCF	Festuca rubra 48
BDF	Agrostis stolonifera 92
BF	Puccinellia maritima 60
	Puccinellia phryganodes (N) 60
D	*Puccinellia rupestris* 60
DF	*Puccinellia rupestris* 60
F	Puccinellia distans 60
	Puccinellia rupestris 60
	Alopecurus bulbosus 100

Vegetative Key to Grasses (Gramineae)

In water

B	*Scolochloa festucacea* (N) 68
	Glyceria plicata 68
BCE	Phragmites australis 102
BD	Glyceria maxima 68
BDE	Catabrosa aquatica 62
BE	Phalaris arundinacea 102
D	Glyceria fluitans 68

Riversides, marshes, fens, dune slacks

ABF	Alopecurus aequalis 100
B	Alopecurus geniculatus 100
BCE	Phragmites australis 102
BCH	Elymus repens 78
BD	Glyceria maxima 68
BDE	Catabrosa aquatica 64
BDF	Poa trivialis 56
	Agrostis canina 92
	Agrostis stolonifera 92
BE	Phalaris arundinacea 102
	Calamagrostis purpurea (N) 96
BEF	*Arctophila fulva* (N) 62
BEG	Calamagrostis canescens 96
BG	*Leersia oryzoides* 112
C	*Sesleria caerulea* (S) 66
CEG	Molinia caerulea ssp. arundinacea 106
D	Glyceria declinata 68
	Glyceria fluitans 68
	Glyceria × pedicellata 68
	Hierochloë odorata 90
	Calamagrostis pseudophragmites (S) 96
DF	Glyceria declinata 68
	Deschampsia cespitosa 88
E	Calamagrostis epigeios 96
	Calamagrostis pseudophragmites (S) 96
EG	*Calamagrostis chalybaea* (N) 96
F	Poa palustris 56
FG	*Calamagrostis stricta* 96
	Trisetum subalpestre (N) 86
G	*Calamagrostis arundinacea* (S) 96

Bare damp ground

ABF	Alopecurus aequalis 100
ACF	*Parapholis incurva* 102
ADF	*Agrostis clavata* (N) 92
	Polypogon maritimus (S) 94
	Milium vernale (S) 104
AF	*Coleanthus subtilis* (S) 102
B	Alopecurus geniculatus 100
BDE	Catabrosa aquatica 64
BDF	Agrostis stolonifera 92
D	Glyceria declinata 68
DF	*Phippsia algida* (N) 60
	Phippsia concinna (N) 60
G	*Elymus alaskanus* (N) 78

Damp meadows

—	Phleum pratense 98
	Alopecurus pratensis 100
ADF	*Polypogon maritimus* (S) 94
AEFG	Bromus hordeaceus 76
AF	*Alopecurus rendlei* (S) 100
	Crypsis alopecuroides (S) 108
BCF	Festuca rubra 48
BD	Glyceria maxima 68
BDF	Poa trivialis 56
	Agrostis canina 92
	Agrostis stolonifera 92
BF	Poa pratensis 56
	Poa subcaerulea 56
	Briza media 66
	Agrostis capillaris 92
BG	*Leersia oryzoides* 112
C	*Sesleria caerulea* (S) 66
CEF	Danthonia decumbens 106
CEG	Molinia caerulea ssp. arundinacea 106
CF	Danthonia decumbens 106
D	*Hierochloë odorata* 90
DF	Deschampsia cespitosa 88
	Deschampsia media (S) 88
DFG	Avenula pubescens 84
E	Anthoxanthum odoratum 90
	Dactylis glomerata 62
EFG	Trisetum flavescens 86
	Anthoxanthum odoratum 90

Vegetative Key to Grasses (Gramineae)

EG	Anthoxanthum odoratum 90
F	Cynosurus cristatus 62
FGH	Hordeum secalinum 82
G	Holcus lanatus 90
	Calamagrostis arundinacea (S) 96
H	Festuca pratensis 46
	Festuca arundinacea 46
	×Festulolium loliaceum 50
	Lolium perenne 50

Arable land

A	*Lolium rigidum* 50
	Avena nuda 84
	Avena brevis 84
	Digitaria ischaemum 110
AD	*Cynosurus echinatus* 62
	Briza minor 66
ADE	*Apera spica-venti* 64
ADEF	*Anthoxanthum aristatum* 90
ADEFG	*Apera interrupta* 64
ADF	Poa annua 54
	Gastridium ventricosum 94
AE	Bromus secalinus 74
	Zea mays 110
	Alopecurus myosuroides 100
AEG	*Bromus commutatus* 74
	Bromus racemosus 74
AEH	*Lolium temulentum* 50
AF	*Avena brevis* 84
AFG	Vulpia myuros 52
AG	Avena strigosa 84
	Avena fatua 84
	Avena sativa 84
AGH	*Triticum spelta* 80
	Triticum aestivum 80
AH	Secale cereale 80
	Hordeum distichon 82
	Hordeum vulgare 82
BCH	Elymus repens 78
BDH	Agrostis stolonifera 92
BE	Agrostis gigantea 92
BG	Holcus lanatus 90

Urban sites, disturbed ground, paths

—	Arrhenatherum elatius 84
A	*Briza maxima* 66
ABG	*Digitaria sanguinalis* 110
AC	*Echinochloa crus-galli* 112
	Setaria spp. 108
ACFG	Vulpia bromoides 52
AD	*Cynosurus echinatus* 62
	Polypogon monspeliensis 94
ADE	*Phalaris canariensis* 102
	Phalaris minor 102
ADEF	*Anthoxanthum aristatum* 90
ADEFG	Bromus sterilis 70
ADEG	Bromus sterilis 70
ADF	Poa annua 54
ADG	*Ventenata dubia* (S) 84
AE	Desmazeria rigida 54
	Zea mays 110
AEFG	*Vulpia unilateralis* 52
	Bromus tectorum 70
	Bromus hordeaceus 76
AEG	*Bromus diandrus* 70
	Bromus rigidus 70
AF	*Poa infirma* 54
	Hordeum murinum 82
AFG	Vulpia myuros 52
	Eragrostis spp. (S) 106
AFGH	Hordeum murinum 82
AG	Avena sativa 84
AGH	Triticum aestivum 80
AH	Lolium multiflorum 50
	Hordeum distichon 82
	Hordeum vulgare 82
BCEF	*Cynodon dactylon* (S) 110
BCH	Elymus repens 78
BDF	Agrostis stolonifera 92
BF	Poa pratensis 56
	Poa compressa 54
D	Arrhenatherum elatius 84
FGH	Hordeum jubatum 82
G	Holcus lanatus 90
H	Lolium perenne 50

Agricultural grassland, rough grassland, road verges

—	Phleum pratense 98
	Alopecurus pratensis 100
	Arrhenatherum elatius 84
ACFG	Vulpia bromoides 52
ADF	Poa annua 54
AEFG	Bromus hordeaceus 76

Vegetative Key to Grasses (Gramineae)

	Bromus racemosus 74
AEG	*Bromus commutatus* 74
AF	*Alopecurus rendlei* (S) 100
AGH	Hordeum murinum 82
AH	Lolium multiflorum 50
BCF	Festuca rubra 48
BCH	Elymus repens 78
BD	Poa trivialis 56
BDF	Poa trivialis 56
	Agrostis stolonifera 92
BF	Poa pratensis 56
	Agrostis capillaris 92
BFG	*Koeleria pyramidata* (S) 86
CF	Festuca nigrescens 48
D	Arrhenatherum elatius 84
DF	Deschampsia cespitosa 88
E	Dactylis glomerata 62
	Calamagrostis epigeios 96
EFG	Trisetum flavescens 86
F	Cynosurus cristatus 62
FGH	Hordeum secalinum 82
G	Holcus lanatus 90
H	Festuca pratensis 46
	Festuca arundinacea 46
	×Festulolium loliaceum 50
	×*Festulolium braunii* 50
	Lolium perenne 50
	Lolium multiflorum 50

Chalk and limestone grassland

—	Arrhenatherum elatius 84
ADF	*Gastridium ventricosum*
AE	Desmazeria rigida 54
AEFG	Vulpia unilateralis 52
B	Brachypodium pinnatum 72
BCF	Festuca rubra 48
BE	Brachypodium pinnatum 72
BF	Poa angustifolia 56
	Briza media 66
	Brachypodium pinnatum 72
	Agrostis capillaris 92
BFG	Koeleria macrantha 86
	Koeleria pyramidata (S) 86
C	Sesleria albicans 66
CF	Sesleria albicans 66
	Festuca amethystina (S) 48
CFG	*Koeleria vallesiana* 86
CFH	Festuca ovina 48

D	Arrhenatherum elatius 84
DF	Avenula pubescens 84
	Avenula pratensis 84
DFG	Avenula pubescens 84
DG	Avenula pratensis 84
E	Anthoxanthum odoratum 90
	Dactylis glomerata 62
EF	Bromus erectus 72
EFG	Trisetum flavescens 86
	Anthoxanthum odoratum 90
	Stipa pulcherrima (S) 104
EG	Anthoxanthum odoratum 90
F	Cynosurus cristatus 62
	Phleum phleoides 98
FG	*Stipa* spp. (S) 104
G	Holcus lanatus 90
	Brachypodium sylvaticum 72
H	Festuca arundinacea 46

Dry sandy grassland and steppe

ACFG	*Micropyrum tenellum* (S) 52
	Vulpia ciliata 52
ADE	*Phalaris minor* 102
ADF	Aira praecox 88
	Aira caryophyllea 86
AE	Desmazeria rigida 54
AEFG	*Vulpia unilateralis* 52
	Bromus madritensis 68
	Bromus hordeaceus 76
AF	*Poa infirma* 54
AFG	Vulpia myuros 52
BCEF	*Cynodon dactylon* (S) 110
BCF	Festuca rubra 48
BF	Poa pratensis 56
	Poa angustifolia 56
	Poa compressa 54
	Agrostis capillaris 92
BFG	Koeleria macrantha 86
BG	Holcus mollis 90
CEF	Danthonia decumbens 106
CF	Danthonia decumbens 106
CFH	Festuca tenuifolia 48
	Festuca ovina 48
	Festuca longifolia 218
	Festuca valesiaca (S) 218
DF	*Poa bulbosa* 54
	Avenula pratensis 84

Vegetative Key to Grasses (Gramineae)

DG Avenula pratensis 84
E Anthoxanthum odoratum 90
EF *Poa badensis* (S) 58
EFG Anthoxanthum odoratum 90
F Cynosurus cristatus 62
Phleum phleoides 98
FG Koeleria glauca 86
Stipa spp. (S) 104

Deciduous woods and hedges
— Poa chaixii 56
Elymus caninus 78
Milium effusum 104
BCG Melica uniflora 66
BDF Poa trivialis 56
BFG *Koeleria pyramidata* (S) 86
BG Holcus mollis 90
CF *Festuca heterophylla* 48
Festuca nigrescens 48
Poa nemoralis 58
DEH *Bromus benekenii* 72
E Dactylis glomerata 62
Calamagrostis epigeios 96
EH Bromus ramosus 72
F Hordelymus europaeus 80
G Brachypodium sylvaticum 72
Elymus caninus 70
Elymus mutabilis (N) 70
H Festuca gigantea 46

Dry woods
BG Holcus mollis 90
CF *Melica picta* (S) 66
D *Hierochloë australis* (S) 90
DF *Deschampsia flexuosa* 88
E *Melica transsilvanica* (S) 66
G Brachypodium sylvaticum 72

Wet woods
B *Poa remota* (S) 56
BCE Phragmites australis 102
BD Poa trivialis 56
Glyceria lithuanica (N) 68
BDF Poa trivialis 56
Agrostis stolonifera 92
BE Phalaris arundinacea 102

BEG Calamagrostis canescens 96
CEG Molinia caerulea 106
D *Hierochloë odorata* 90
DF Deschampsia cespitosa 88
EG *Cinna latifolia* (N) 64
Calamagrostis chalybaea (N) 96
G *Calamagrostis arundinacea* (S) 96

Rocky woods
— *Poa chaixii* 56
B *Poa remota* (S) 56
BC Melica nutans 66
BCF Melica nutans 66
BE *Calamagrostis villosa* (S) 96
BEG *Calamagrostis lapponica* (N) 96
BF Agrostis capillaris 92
CE *Calamagrostis varia* (S) 96
CF Poa nemoralis 58
DF Deschampsia flexuosa 88
E *Festuca altissima* 46
Anthoxanthum odoratum 90
EFG Anthoxanthum odoratum 90
EG Anthoxanthum odoratum 90
F *Agrostis mertensii* (N) 92

Conifer woods
BC Melica nutans 66
BCF Melica nutans 66
BE *Calamagrostis villosa* (S) 96
BEG *Calamagrostis lapponica* (N) 96
BFG Koeleria macrantha 86
CF Poa nemoralis 58
DF Deschampsia flexuosa 88
EH Bromus ramosus 72

Upland grassland
BCF Festuca rubra 48
BDF *Poa arctica* (N) 58
BF *Poa alpigena* (N) 56
Poa subcaerulea 56
Poa angustifolia 56
Agrostis capillaris 92
C Sesleria albicans 66

Vegetative Key to Grasses (Gramineae)

CE   *Calamagrostis varia* (S)
     96
CEF   Danthonia decumbens
     106
CF   Sesleria albicans 66
     Danthonia decumbens
     106
CFH   Festuca tenuifolia 48
     Festuca ovina 48
DF   *Poa arctica* (N) 58
F   *Poa arctica* (N) 58
     Cynosurus cristatus 62
     Nardus stricta 106
G   Holcus lanatus 90

Heaths and moors
ADF   Aira caryophyllea 88
     Aira praecox 88
BDF   *Poa arctica* (N) 58
BE   *Calamagrostis villosa* (S)
     96
BEG   *Calamagrostis lapponica*
     (N) 96
BF   *Poa alpigena* (N) 56
     Agrostis capillaris 92
CEF   Danthonia decumbens
     106
CEG   Molinia caerulea 106
CF   Danthonia decumbens
     106
CFH   Festuca ovina 48
     Festuca vivipara 48
     Festuca tenuifolia 48
DEF   Agrostis vinealis 92
DF   Deschampsia flexuosa 88
     Deschampsia setacea 88
     Agrostis curtisii 92
     Poa arctica (N) 58
F   Nardus stricta 106
     Poa arctica (N) 58

Mountain flushes
—   *Phleum alpinum* 98
     Alopecurus alpinus 100
BDF   *Poa supina* (S) 54
F   *Alopecurus alpinus* 100

Bogs, tundra
BEG   *Calamagrostis lapponica*
     (N) 96
CEG   Molinia caerulea 106
DF   *Deschampsia setacea* 88
     Hierochloë alpina (N) 90
FG   *Calamagrostis scotica* 96

Mountain rocks
—   *Alopecurus alpinus* 100
BDF   *Poa supina* (S) 54
BF   *Poa alpina* 58
CF   Poa nemoralis 58
     *Achnatherum
     calamagrostis* (S) 104
CFH   Festuca ovina 48
     Festuca vivipara 48
DF   *Poa laxa* 58
     Poa alpina 58
     Poa ×jemtlandica 58
F   *Poa flexuosa* 58
     Poa glauca 58
     Alopecurus alpinus 100
FG   *Trisetum spicatum* (N)
     86

Mountain tops
(excluding flushes)
BCF   *Festuca richardsonii* 48
BF   Agrostis capillaris 92
CFH   Festuca ovina 48
DF   Deschampsia flexuosa 88
     Hierochloë alpina (N) 90

Walls, lowland rocks
ADF   Aira praecox 88
AE   Desmazeria rigida 54
AEFG   Bromus hordeaceus 76
BF   Poa pratensis 56
     Poa compressa 54
CFG   *Koeleria vallesiana* 86
DF   *Poa bulbosa* 54
E   Dactylis glomerata 62
EF   *Melica ciliata* (S) 66

Key to Sedges (Cyperaceae)

State 1

State 2

A Plant forming conspicuous raised tussocks

a plant not forming raised tussocks

B Shoots isolated or in patches

b shoots in tufts or small tussocks

C Stems leafless, green

c obvious leaves present at least at base

D Leaves 3 mm or more wide

d leaves less than 3 mm wide

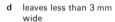

E Lower bract at least as long as inflorescence

e lower bract shorter than inflorescence

F Flowers hermaphrodite

f flowers unisexual

G Spikes all similar, or only 1

g spikes more than 1, dissimilar

Key to Sedges (Cyperaceae)

H Stigmas 3   **h** stigmas 2

For characters I, J and K ignore lowest spike

I Spikes overlapping or clustered, or lowest distant

i spikes spaced or only 1

J Spikes oblong or cylindrical (at least 3 times as long as wide)

j spikes ovoid or globular (less than 2 times as long as wide)

K Spikes stalked (excluding terminal)

k spikes unstalked or only 1, or very short-stalked

L Female glumes awned or acuminate with a ' drawn-out point

l female glumes acute or obtuse and not with a drawn-out point

M Female glumes dark or red-brown

m female glumes pale brown or greenish, except for midrib

N Fruit not or shortly beaked

n fruit with a long beak

O Fruit hairy, downy or rough

o fruit entirely smooth

ABEIJKMN	Carex nigra 164	
ABEIJMN	Carex nigra 164	
ABIJMN	Carex nigra 164	
ACFGHIN	Scirpus cespitosus 118	
ADGIKM	Carex paniculata 128	
ADGIM	Carex paniculata 128	

ADIJM	Carex elata 164
AEIJKMN	Carex nigra 164
AFGHIMN	Eriophorum brachyantherum 120
AFGHIN	Eriophorum vaginatum 120

33

Key to Sedges (Cyperaceae)

Key to Sedges (Cyperaceae)

BGIN • •utricles red-brown, faintly veined: Carex praecox 132

BGIN Carex chordorrhiza 134
BGJM Carex pulicaris 166
BHIJKLMO Carex flacca 148
BHIJLMO Carex caryophyllea 154
BHIK Carex limosa 158
BHIKM •female spikes less than 7-flowered: Carex limosa 158
  •female spikes more than 8-flowered: Carex rariflora 158
BHILMO Carex caryophyllea 154
BHIMN Carex holostoma 160
BHJKLO Carex hirta 140
BHJKM Carex panicea 148
BHJM Carex rostrata 144
BHJMN Carex rotundata 140
BHKM •spikes erect, leaves more than 2 mm wide: Carex hostiana 150
  •spikes nodding, leaves less than 2 mm wide: Carex laxa 158
BIJM Carex bigelowii 164
BIJMN Carex nigra 164
BJMN Carex saxatilis 140

CDEFGHIMN Schoenus nigricans 126
CDFGHIMN Schoenus nigricans 126
CEFGHILMN Scirpus mucronatus 116
CEFGHIMN Schoenus nigricans 126
CFGHIM Eleocharis multicaulis 122
CFGHIMN •slender annual: Scirpus pumilus 118
  •robust perennial: Schoenus nigricans 126
CFGHIN •glumes as long as spikelet▷
  • •stems 3-angled: Scirpus hudsonianus 118
  • •stems round: Scirpus cespitosus 118
  •glumes shorter than spikelet▷
  • •stems less than 0.5 mm diameter: Eleocharis parvula 122
  • •stems more than 0.5 mm diameter: Eleocharis quinqueflora 122

CFGI •annual, stems c. 1 mm diameter: Eleocharis ovata 122
  •perennial, stems more than 1 mm diameter: Eleocharis palustris agg. 122
CFGIM Eleocharis palustris agg. 122

DEFGHIMN •lower bract much longer than inflorescence: Schoenus nigricans 126
  •lower bract equalling inflorescence: Schoenus ferrugineus 126

DEHIJKL Carex pseudocyperus 142
DEHIJKLM Carex riparia 142
DEHIJKLMO Carex flacca 148
DEHIJKLN Carex pallescens 148
DEHIJL Carex acutiformis 142
DEHIJLM Carex melanostachya 142
DEHIK Carex flava 152
DEHIKLN Carex pallescens 148
DEHIKM •female glumes blackish: Carex atrata 160
  •female glumes brownish: Carex flava agg. 152
DEHIM Carex flava agg. 152
DEHJK •lowest spike drooping: Carex sylvatica 146
  •lowest spike erect▷
  • •1 male spike: Carex punctata 150
  • •more than 1 male spike: Carex hordeistichos 140
DEIJKLM •large plant, leaves more than 4 mm wide: Carex recta 162
  •smaller, leaves less than 4 mm wide: Carex vacillans 162
DEIJKMN •underside of leaves greyish: Carex acuta 164
  •underside of leaves green: Carex aquatilis 162
DEIJLM Carex acutiformis 142
DEIJMN Carex acuta 164
DFGHIJN Scirpus radicans 114
DFGHIKMN Eriophorum latifolium 120

35

Key to Sedges (Cyperaceae)

DFGHIMNPQR   Schoenus ferrugineus 126

DGI   •dry soils: Carex divulsa ssp. leersii 130
     •wet soils: Carex elongata 130

DGIJ   Carex elongata 130

DGIJM   Carex elongata 130

DGIL   •stout, thick-stemmed: Carex otrubae 130
     •slender, thin-stemmed: Carex divulsa ssp. leersii 130

DGILM   •stem triangular, winged: Carex vulpina 130
     •stem unwinged▷
     ••stem-base brown: Carex muricata 130
     ••stem-base purplish-red: Carex spicata 130

DGILMN   Carex muricata 130

DGIM   •dry soils: Carex muricata 130
     •wet soils: Carex elongata 130

DGIMN   Carex muricata 130

DGL   Carex otrubae 130

DHIJKLM   Carex buxbaumii 160

DHIJKM   •large, spikes drooping, woods: Carex pendula 144
     •smaller, spikes erect, bogs: Carex livida 148

DHIJKMNO   Carex digitata 154

DHILM   Carex fritschii 156

DHIMNO   Carex ericetorum 154

DHJK   •lowest spike erect: Carex punctata 150
     •lowest spike drooping: Carex sylvatica 146

DHJKL   •leaves more than 5 mm wide: Carex laevigata 148
     •leaves less than 6 mm wide: Carex distans 150

DHJKLM   •leaves more than 5 mm wide: Carex laevigata 148
     •leaves less than 6 mm wide▷
     ••spikes erect: Carex binervis 150
     ••spikes nodding: Carex atrofusca 158

DHJKM   •lowest bract with inflated sheath: Carex vaginata 148
     •lowest bract scarcely sheathing: Carex panicea 148

DHJKN   Carex strigosa 146

DHKLM   •terminal spike male at top: Carex atrofusca 158
     •terminal spike female at top: Carex fuliginosa 158

DHKM   Carex depauperata 146

DIJKMN   Carex acuta 164

DIJM   Carex elata 164

DIJMN   Carex acuta 164

EFGH   Rhynchospora alba 126

EFGHIKLN   Scirpus holoschoenus 114

EFGHIKMN   Cyperus fuscus 124

EFGHIMN   •bract short: Scirpus setaceus 118
     •bract long: Scirpus supinus 116

EFGI   Rhynchospora alba 126

EFGIKN   Cyperus flavescens 124

EFGIN   Blysmus compressus 118

EG   Carex remota 136

EGI   Carex bohemica 136

EGILM   Carex divisa 134

EHI   Carex flava agg. 152

EHIJK   Carex capillaris 146

EHIJKLN   Carex pallescens 148

EHIK   Carex flava agg. 152

EHIKLM   Carex extensa 152

EHIKLN   Carex pallescens 148

EHIKM   Carex flava agg. 152

EHIM   Carex norvegica 160

EHJK   Carex secalina 140

EHJKMO   Carex lasiocarpa 140

EHKLM   Carex extensa 152

EIJKMN   Carex nigra 164

EIJMN   Carex nigra 164

EIKM   Carex bicolor 160

EIMN   Carex rufina 160

FGH   Rhynchospora alba 126

FGHILN   Scirpus setaceus 118

FGHIMN   Eriophorum brachyantherum 120

FGHIN   Eriophorum vaginatum 120

FGI   Rhynchospora alba 126

G   •single spike: Carex capitata 166
     •several spikes▷

Key to Sedges (Cyperaceae)

        ••spikes star-shaped:
Carex echinata 136

        ••spikes not star-
shaped▷

        •••leaves more than 2
mm wide: Carex
divulsa 130

        •••leaves less than 2 mm
wide: Carex loliacea
138

GH Carex pauciflora 166
GHIJM Kobresia myosuroides 126
GHIM Kobresia simpliciuscula
126
GHJM Carex rupestris 166
GHJMO Carex scirpoidea 136
GI •beak of fruit bifid▷
        ••fruit winged: Carex
ovalis 134
        ••fruit unwinged▷
        •••tall, fens: Carex
diandra 128
        •••short, coastal sands:
Carex maritima 134
        •beak of fruit not bifid▷
        ••spikes dark brown:
Carex heleonastes 138
        ••spikes pale brown▷
        •••less than 2 mm wide:
Carex mackenziei 138
        •••leaves 2–3 mm wide▷
        ••••bogs: Carex curta
curta 138
        ••••saltmarshes: Carex
mackenziei 138
GIJ Carex elongata 130
GIJM Carex elongata 130
GIKM Carex appropinquata 128
GIL Carex ovalis 134
GILM Carex spicata 130
GILN Carex muricata ssp.
lamprocarpa 130
GIM •fruit winged: Carex
macloviana 134
        •fruit unwinged▷
        ••fruit more than 3 mm:
Carex elongata 130
        ••fruit less than 3 mm▷
        •••leaves less than 2 mm
wide: Carex lachenalii
138
        •••leaves more than 2
mm wide: Carex
brunnescens 138
GIMN Carex glareosa 138
GIN Carex tenuiflora 138

GJM Carex pulicaris 166
GK Carex divulsa 130
GM Carex nardina 166

HIJKLMNO Carex pediformis 154
HIJKMNO Carex digitata 154
HIJKMO Carex pediformis 154
HIJLM Carex melanostachya 142
HIJLMNO Carex caryophyllea 154
HIJO Carex ornithopoda 154
HIKLM Carex buxbaumii 160
HIKLMO Carex umbrosa 154
HIKM •densely tufted, leaves
less than 1 mm wide:
Carex glacialis 154
        •loosely tufted, leaves
more than 1 mm wide:
Carex liparocarpos 156
HILMNO Carex hallerana 154
HIMO Carex pilulifera 156
HIM •plant less than 20 cm,
fruit beak bifid: Carex
supina 156
        •plant more than 20 cm,
fruit beak truncate:
Carex stylosa 160
HIMNO •fruit blackish: Carex
montana 156
        •fruit green: Carex
ericetorum 154
HIO Carex ornithopoda 154
HJKLM Carex binervis 150
HJKM Carex panicea 148
HJKMO Carex lasiocarpa 140
HJMNO Carex humilis 154
HK •female glumes very
pale, C Europe: Carex
alba 156
        •female glumes pale
brown, W Europe:
Carex punctata 150
HKL Carex distans 150
HKLM •basal leaf sheath pale:
Carex fuliginosa 158
        •basal leaf sheath dark:
Carex sempervirens
158
HKLMO Carex sempervirens 158
HKM Carex depauperata 146
HM Carex supina 156
HMO Carex globularis 156

IJ Carex cespitosa 164

37

Key to Rushes and Wood-rushes (Juncaceae)

State 1

A Stem less than 30 cm tall

B No leaves; inflorescence apparently lateral

C Leaves (or stems if B) hairy

D Leaves (or stems if B) cylindrical or thread-like

E Flowers in tight heads or spikes, or solitary and terminal

F Tepals dark or reddish brown

G Tepals less than 3 mm

H Tepals blunt-tipped

I Tepals of different sizes

J Capsule shorter than tepals

State 2

a stem more than 30 cm tall

b leaves present; inflorescence terminal

c leaves hairless

d leaves flat

e flowers in loose branching clusters

f tepals pale brown or greenish

g tepals more than 3 mm

h tepals pointed

i tepals more or less the same size

j capsule longer than tepals

Key to Rushes and Wood-rushes (Juncaceae)

K Capsule blunt-tipped

k capsule pointed

ABEFHI	Juncus arcticus 170
ABEGI	Juncus filiformis 170
ABEGIK	Juncus filiformis 170
ABH	Juncus balticus 170
AC	Luzula luzulina 182
ACEFGIK	Luzula sudetica 180
ACEFGJK	•flowers in several dense clusters: Luzula multiflora 180
	•flowers in a single spike: Luzula spicata 180
ACEFGK	•flowers in several dense clusters: Luzula multiflora 180
	•flowers in a single dense cluster: Luzula arctica 180
	•flowers in a single spike: Luzula spicata 180
ACEFJK	•anthers scarcely longer than filaments: Luzula multiflora 180
	•anthers much longer than filaments: Luzula campestris 180
ACEFK	Luzula multiflora 180
ACEGIJK	Luzula pallescens 180
ACFG	Luzula wahlenbergii 182
ACFJ	Luzula forsteri 182
ACFK	Luzula pilosa 182
ADEF	Juncus canadensis 178
ADEFG	Juncus trifidus 172
ADEFGHK	Juncus biglumis 176
ADEFHJ	Juncus pygmaeus 176
ADEFJ	Juncus pygmaeus 176
ADEGHK	Juncus compressus 168
ADEHJ	Juncus pygmaeus 176
ADEHK	•bracts short, broad: Juncus triglumis 176
	•lowest bract leaf-like: Juncus stygius 176
ADEJ	Juncus pygmaeus 176
ADFG	•leaves thread-like, plant branched: Juncus bulbosus 172
	•leaves tubular, plant unbranched: Juncus articulatus 178
ADFGH	Juncus anceps 178

ADFGHK	Juncus alpinus 178
ADFH	Juncus squarrosus 174
ADFHK	Juncus gerardi 168
ADG	Juncus bulbosus 172
ADGHK	Juncus compressus 168
AEFG	Juncus planifolius 174
AEFGK	•leaves less than 1 mm wide, hairless: Juncus tenageia 172
	•leaves more than 1 mm wide △
	••flower-stalks drooping: Luzula arcuata 180
	••flower-stalks straight: Luzula confusa 180
AEFIJ	Juncus capitatus 176
AEFIK	Juncus castaneus 176
AEFK	Juncus castaneus 176
AEIJ	Juncus capitatus 176
AFG	Luzula alpinopilosa 182
AFGJ	Luzula alpinopilosa 182
AFGK	Juncus tenageia 172
AI	Juncus bufonius 172
AIJ	Juncus bufonius 172
AIJK	Juncus sphaerocarpus 172
AJ	•leaves as long as stems: Juncus tenuis 174
	•leaves one-third as long as stems: Juncus dudleyi 174
BDEF	Juncus acutus 168
BDEFH	Juncus acutus 168
BDEFHI	Juncus acutus 168
BDEFI	Juncus acutus 168
BDEI	Juncus maritimus 168
BEFHI	Juncus arcticus 170
BEGI	Juncus inflexus 170
BEGJ	•stem finely ridged: Juncus effusus 170
	•stem strongly ridged: Juncus conglomeratus 170
BEGJK	Juncus conglomeratus 170
BEI	Juncus inflexus 170
BEJ	Juncus conglomeratus 170

Key to Rushes and Wood-rushes (Juncaceae)

BEJK	Juncus conglomeratus 170
BGJ	Juncus effusus 170
BH	Juncus balticus 170
CEFGJK	Luzula multiflora 180
CEFGK	Luzula multiflora 180
CEFJK	Luzula multiflora 180
CEFK	Luzula multiflora 180
CEIJK	Luzula nivea 182
CFGJ	Luzula parviflora 182
CFGK	Luzula desvauxii 182
CFI	Luzula sylvatica 182
CFK	Luzula pilosa 182
CGI	Luzula luzuloides 182
CI	Luzula luzuloides 182
DE	Juncus ensifolius 178
DEF	Juncus canadensis 178
DEI	Juncus maritimus 178
DFG	Juncus articulatus 176
DFGH	Juncus anceps 178
DFGHK	Juncus alpinus 178

DFGI	•flowers blackish-brown: Juncus atratus 178
	•flowers mid-brown: Juncus articulatus 178
DFH	J. squarrosus 174
DFHK	Juncus gerardi 168
DFI	•leaves angled in section: Juncus atratus 178
	•leaves round in section: Juncus acutiflorus 178
DG	Juncus subulatus 168
DGH	Juncus subnodulosus 178
DI	Juncus maritimus 168
FG	Luzula alpinopilosa 182
FGJ	Luzula alpinopilosa 182
J	•leaves as long as stems: Juncus tenuis 174
	•leaves one-third as long as stems: Juncus dudleyi 174

Keys to Pinnate Ferns (Filicopsida)

State 1   **State 2**

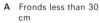

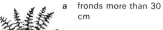

A Fronds less than 30 cm   **a** fronds more than 30 cm

B Fronds isolated   **b** fronds forming a tuft or crown

C Leaf-stalk scaly   **c** smooth or with a few scales at base

D Leaf-stalk less than one-quarter the length of the blade   **d** stalk more than one-quarter of blade

E Leaf-stalk brown or black   **e** stalk pale or green except at base

F Sterile and fertile fronds morphologically distinct   **f** all fronds similar

G Spore-cases on leaf-margins or at leaf bases   **g** spore-cases on midrib, veins (or none)

H Spore-cases linear, elliptical etc. but not round   **h** spore-cases round (or none)

Note: Five species of fern described in this book have unlobed leaves: *Ophioglossum lusitanicum, O. azoricum, O. vulgatum, Botrychium simplex* (p. 196) and *Phyllitis scolopendrium* (p. 198).
  One species has forked fronds: *Asplenium septentrionale* (p. 200).

Key to Pinnate Ferns (Filicopsida)

Key to 1-pinnate Ferns

A   • leaves pinnate, with lobed pinnae: Dryopteris cristata 210
    • leaves pinnately lobed: Polypodium vulgare 214

ABDEG   • leaf-lobes flat: Hymenophyllum tunbrigense 196
    • leaf-lobes bent back: Hymenophyllum wilsonii 196

ABDF   Botrychium lunaria 194

ABF   Botrychium simplex and B. lunaria 194

ACD   Polystichum lonchitis 208

ACDEFGH   Blechnum spicant 214

ACDH   Ceterach officinarum 202

ACEG   Woodsia ilvensis 206

ADEH   Asplenium trichomanes 200

AEG   Woodsia alpina 206

AEH   Asplenium adulterinum 200

AG   Woodsia glabella 206

AH   • spore-cases linear▷
    •• main midrib winged, coastal: Asplenium marinum 200
    •• main midrib unwinged, mountains: Asplenium viride 200
    • spore-cases elliptical▷
    •• leaf-blade narrow, widest near middle; pinnae entire: Polypodium interjectum 214
    •• leaf-blade broader, widest near base; pinnae serrate: Polypodium australe 214

B   Thelypteris palustris 198

BC   Phegopteris connectilis 198

BCDEGH   Pteris vittata 219

BCEGH   Pteris cretica 219

BF   Onoclea sensibilis 204

BFH   Onoclea sensibilis 204

C   Polystichum spp. (except P. lonchitis) 208

CD   Polystichum lonchitis 208

CDEFGH   Blechnum spicant 214

DFGH   Matteuccia struthiopteris 204

F   Dryopteris cristata 210

Key to 2-pinnate Ferns

A   Cystopteris fragilis (and C. dickieana) 206

AB   • fronds oblong in outline: Thelypteris palustris 198
    • fronds triangular in outline▷
    •• bright, pale green, soft in texture: Gymnocarpium dryopteris 214
    •• dull green, tough in texture: Gymnocarpium robertianum 214

ABC   Phegopteris connectilis 198

ABDEG   leaf-lobes flat: Hymenophyllum tunbrigense 196
    leaf-lobes bent-back: Hymenophyllum wilsonii 196

ABDF   Botrychium spp. (B. boreale, B. lanceolatum, B. matricariifolium) 194

ABF   Botrychium multifidum 194

AC   leaves less than 20 cm, Arctic: Dryopteris fragrans 212
    leaves more than 20 cm, limestone: Dryopteris submontana 210

ACEG   Woodsia ilvensis 206

ADE   Asplenium billotii 202

AE   Cystopteris fragilis (and C. dickieana) 206

AEF   Anogramma leptophylla 196

AEG   Woodsia alpina 206

AEGH   Adiantum capillus-veneris 196

AEH   • many leaflets: Asplenium adiantum-nigrum, A. onopteris 202
    • few leaflets: Asplenium ruta-muraria 200

AG   Woodsia glabella 206

Keys to Pinnate Ferns (Filicopsida)

AH • lowest leaf-segments shortest: Asplenium fontanum 202
• not shortest: Asplenium forisiense 202

B Thelypteris palustris 198

BC Phegopteris connectilis 198

BF Onoclea sensibilis 204

BFH Onoclea sensibilis 204

C • leaves with hair-like teeth: Polystichum spp. 208
• leaves with pointed but not hair-like teeth: Dryopteris spp. 210–212

CD • leaves leathery, leaflets hair-pointed: Polystichum aculeatum 208
• leaves soft, not hair-pointed: Oreopteris limbosperma 198

CH Athyrium filix-femina 204

DF Osmunda regalis 192

DFGH Matteuccia struthiopteris 204

EGH Adiantum capillus-veneris 196

EH Asplenium adiantum-nigrum, A. onopteris 202

F • tall, mound-building: Osmunda regalis 204
• shorter, crown-forming: Dryopteris cristata 210

Key to 3- or 4-pinnate Ferns

A Cystopteris fragilis (and C. dickieana) 206

ABC • lowest, downward pointing segment longest: Cystopteris montana 206
• not longest: Cystopteris sudetica 206

ABDF Botrychium lanceolatum 194

ABF Botrychium multifidum 194

ABFGH Cryptogramma crispa

ABGH Trichomanes speciosum 196

AC see **ABC**

ACD Athyrium distentifolium 204

ACE Dryopteris aemula 212

AEF Anogramma leptophylla 196

AFGH Cryptogramma crispa 196

BC see **ABC**

BCEGH Pteridium aquilinum 192

BCH Diplazium sibiricum 204

BDF Botrychium virginianum 194

BGH Pteridium aquilinum 192

BH Diplazium sibiricum 204

C • robust, with thick petiole: Dryopteris spp. 210–212
• delicate, with slender petiole: see **ABC**

CD Athyrium distentifolium 204

CE Dryopteris aemula 212

CH Athyrium filix-femina 204

EGH Adiantum capillus-veneris 196

EH Asplenium adiantum-nigrum 202

The Grass Family Gramineae

See p. 13 for introduction to grasses.

BAMBOOS *Bambusoideae.* Perennials growing in *clumps* or patches, with hollow *woody* stems (and so used as canes in gardens), sheaths on the main stem and flat leaves. Rarely flowering in Europe, where several species, natives of Asia, are naturalised, usually in or near parks, gardens or waste ground. Few vernacular names are used in Europe, but some Japanese names are given here.

1. **Chimakizasa** *Sasa palmata.* Rhizomes far-creeping, so making extensive patches. Stems erect, to 2 m, usually mottled *purple*, with a *single* branch at each node. Leaves *very large*, to 40 cm long and 90 mm broad, shining green above, paler beneath. B, F. **1a*** *S. veitchii* is shorter, with stems often purplish and glaucous, and smaller, blunter leaves with broad whitish margins, their stalks often purplish.

2. **Himalayan Bamboo** *Arundinaria jaunsarensis (A. anceps).* Rhizomes far-creeping, patch-forming. Stems to 4 m, *arching* when mature, green or brown, with *numerous* branches at each node; sheaths soon falling. Leaves 7–16 cm by 15–20 mm, *shrivelling* in cold winters. B, F. **2a*** *Sinarundinaria nitidia* forms dense clumps and has purplish or greyish stems and dark, usually hairy, more persistent sheaths. **2b*** *Thamnocalamus spathaceus (A. murielae)* has greener stems with hairless sheaths.

3. **Narihiradake** *Semiarundinaria fastuosa.* Rhizomes very shortly creeping, forming clumps. Stems to 8 m, erect, sometimes *flattened* at top, with several short branches at each node; sheaths *hanging* inverted before falling. Leaves 10–15 cm by 15–20 mm. B, F.

4. **Metake** *Pseudosasa japonica.* Rhizomes very short. Stems to 6 m, erect, with a *single* branch at each *upper node only*. Leaves 8–30 cm by 20–35 mm, partly *glaucous* below. B, F.

5. *Pleioblastus simonii.* Rhizomes creeping. Stems to 7 m, *arching* when mature, with several branches at each node. Leaves 5–30 cm by 15–35 mm, usually half *glaucous* below. B, F. **5a*** *P. chino* is shorter, to 3 m, with leaves all green on both sides.

6. *Sasaëlla ramosa (Arundinaria vagans).* Rhizomes far-creeping, forming dense patches. Stems slender, comparatively short, to 1.5 m, slightly *zigzag*, with only *1 branch* at each node on top half of stem only; sheaths persistent, leaves 12–20 cm by 10–25 mm, sometimes with narrow *whitish edges* in hard winters. B, F.

Note: Species of *Phyllostachys*, rarely naturalised in Europe, can be distinguished by their grooved stems.

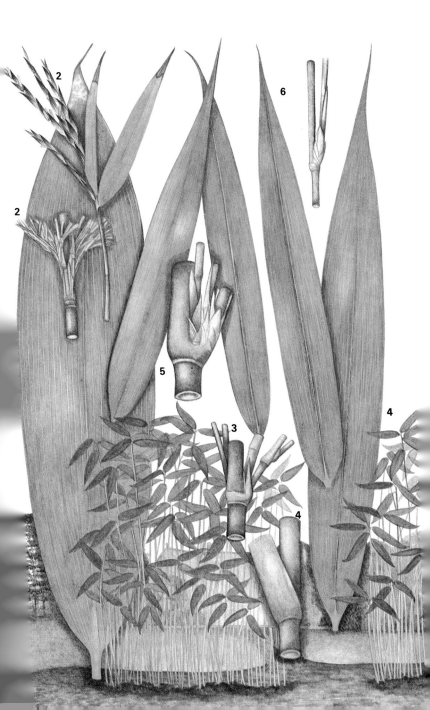

The Grass Family Gramineae

FESCUES *Festuca* (pp. 46–48). Perennials, with the spikelets in panicles.

1. **Wood Fescue** *Festuca altissima.* Tall, to 120 cm, tufted, hairless. Leaves smooth, up to 10 mm wide, with distinct auricles and long (5 mm) ligules. Spikelets △a 5–8 mm, *unawned* and so looking more like *Calamagrostis* than *Festuca*; flowering June–July. Local in moist and shady places, in beech and other woods, on rocks and frequently by streams. T. Map 1.

2. **Giant Fescue** *Festuca gigantea.* Tall, to 150 cm, loosely tufted, hairless. Leaves *shining*, rough-edged, up to 18 mm across, with conspicuous *auricles* and dark *purplish-red* leaf-junctions (cf. Hairy Brome (p. 72) which has green auricles and hairy leaf-sheaths). Panicles drooping, with spikelets △b 8–13 mm, *long-awned*; flowering July–August. Common in woods on most soils and on shady banks. T. Map 2.

3. **Meadow Fescue** *Festuca pratensis.* Medium tall, to 80 cm, loosely tufted, hairless. Leaves narrow but flat, to 4 mm, with green auricles. Panicle branches in pairs, one of which usually has only a *single* spikelet; spikelets △c 11–12 mm, rarely to 15 mm, unawned; flowering June–July. Common but scattered in grassland, especially on heavy and neutral soils. T. Map 3. Hybridises with both Perennial and Italian Rye-grasses (p. 50).

4. **Tall Fescue** *Festuca arundinacea.* Much larger, stouter, taller (to 200 cm) and *more tufted* than Meadow Fescue (3), forming conspicuous tussocks, not unlike Tufted Hair-grass (p. 88), which has rougher, darker green leaves. Leaves up to 10 mm wide, with conspicuous auricles. *All* panicle branches with several spikelets; spikelets △d 10–18 mm, stouter, not or shortly awned; flowering May–July, appearing 3–4 weeks before Meadow Fescue. Common on road verges and in other grassy places. T. Map 4. Hybridises, rarely, with rye-grasses (p. 50).

△ ×3

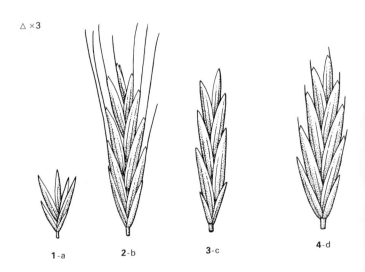

1-a   **2**-b   **3**-c   **4**-d

4   **1**   **3**   **2**

The Grass Family Gramineae

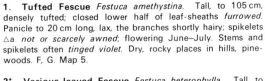

△ ×3

1-a

1. Tufted Fescue *Festuca amethystina*. Tall, to 105 cm, densely tufted; closed lower half of leaf-sheaths *furrowed*. Panicle to 20 cm long, lax, the branches shortly hairy; spikelets △a *not or scarcely awned*; flowering June–July. Stems and spikelets often *tinged violet*. Dry, rocky places in hills, pine-woods. F, G. Map 5.

2.* Various-leaved Fescue *Festuca heterophylla*. Tall, to 150 cm, densely tufted. Basal leaves *thread-like*, lax, *c.* 0.5 mm wide; stem leaves *flat*, 2–3 mm wide, their sheaths *open*. Panicle green, nodding, the spikelets △b shiny, usually shortly awned; flowering June–July. Woodland edges and clearings, mainly on calcareous soils; sown as an ornamental grass. F, G, (B, S). Map 6.

2-b

3.* Red Fescue *Festuca rubra*. Common and very variable, loosely tufted or *shortly creeping*; stems to 60 cm (rarely 110 cm), but usually much shorter; leaf-sheaths *completely closed* when young. Leaves of non-flowering shoots thread-like, of flowering stems *flat*, green, glaucous or bluish. Panicle loose or tight; spikelets △c 7–10 mm, green, glaucous or violet-tinged, awned, sometimes hairy; anthers dull purple; flowering May–June. Grassy places throughout the region, with distinct forms in dunes and saltmarshes. T. Map 7. **3a* Rush-leaved Fescue** *F. juncifolia* has long runners, stiff thick leaves, and lemmas 7–10 mm; mobile dunes. B, F, G. Map 8. **3b* Chewings Fescue** *F. nigrescens* is tufted and has no rhizomes but longer awns and hairless spikelets. Grassland, commonly sown, north to south Sweden. **3c* Arctic Fescue** *F. richardsonii* has tight panicles 4–7 cm, the spikelets densely white-hairy; Arctic, possibly Scotland. **3d Northern Fescue** *F. diffusa* has keeled leaves and spikelets with many florets; north central Europe.

3-c

4.* Sheep's Fescue *Festuca ovina* agg. Common and very variable, but *densely tufted*. Stems to 45 cm (rarely 70 cm), their sheaths *split to more than half-way*; leaves *all* thread-like, green or glaucous. Spikelets △d 5–6 mm, green, glaucous or violet-tinged, awned, usually hairless; flowering June–July. Dry grassland on downs, heaths and moors throughout the region. T. Map 9. More than a dozen very similar species also occur, including **4a* Fine-leaved Sheep's Fescue** *F. tenuifolia*, with even finer leaves and unawned spikelets, on acid soils (map 10); and **4b* Blue Fescue** *F. caesia*, with bluish leaves and sheaths open to base. B, F, G. The rest can only be reliably separated microscopically; see Appendix 1 (p. 218) and maps 11 and 12.

4-d

5.* Viviparous Fescue *Festuca vivipara*. Stems and leaves like Sheep's Fescue (4), Fine-leaved Sheep's Fescue (4a) and other members of the aggregate, but the flowers are replaced by *green plantlets* that sprout tiny leaves and fall off and root. Other grasses do this occasionally, and a few, such as Alpine Meadow-grass (p. 58), habitually, but only this species never has flowers. Moors and mountains. B, S. Map 13.

1

2

5

3

4

The Grass Family Gramineae

△ ×1

1: **Hybrid Fescue** × *Festulolium loliaceum*. The hybrid between Meadow Fescue (p. 46) and Perennial Ryegrass (2), looking very distinct from both parents. Loosely tufted; stems to 120 cm. Leaves *flat or folded* when young, hairless. Inflorescence either a spike with unstalked spikelets or a panicle with spikelets unstalked at the top but stalked lower down. Spikelets △a usually *unawned* or minutely pointed, *sterile*, i.e. with no anthers or stigmas. Grassy places with either Perennial Ryegrass or both parents. T. **1a*** × *F. braunii*, the analogous but rarer hybrid with Italian Ryegrass (3), is very similar, but has the young leaves rolled not folded, and some spikelets shortly awned. Other hybrids, including both ryegrasses with either Tall or Giant Fescues (p. 46), occur rarely.

1-a

RYEGRASSES *Lolium* (2–4), being highly nutritious to livestock, are the most widely sown agricultural grasses, in an enormous range of cultivars, varying greatly in growth habit (prostrate to erect), timing of growth, size and other characters.

2-b

2: **Perennial Ryegrass** *Lolium perenne*. A very common tufted grass; stems to 90 cm. Leaves flat, or folded when young, hairless. Inflorescence a spike, with numerous unstalked *un-awned* spikelets △b, much longer than the glumes; anthers pale yellow; flowering May–November. Grassy places. T. Map 14. The best known of many varieties are S 23 for pasture and S 24 for hay-making. The hybrid with Italian Ryegrass (3), intermediate in all characters, is not uncommon.

3-c

3: **Italian Ryegrass** *Lolium multiflorum*. Tufted annual or biennial, differing from Perennial Ryegrass (2), with which it hybridises, especially in the youngest leaf being *rolled* and its spikelets △c having *long awns*; flowering June–July. Another important fodder plant with many cultivars, widely naturalised on roadsides and field borders. T. Map 15. S 22 is a much used variety and another, Westerwold's (sometimes wrongly called *L. westerwoldicum*), is annual.

4-d

4: **Darnel** *Lolium temulentum*. Once a serious weed of oats and other crops and a candidate for being the biblical 'tare', but now a rare casual. *Annual*, stems to 120 cm. Leaves flat, often glossy beneath. Inflorescence a spike with spikelets △d usually *awned*, the outer glume *equalling or longer than* the whole spikelet, the lemmas *swollen* in fruit; flowering June–August. (T). **4a* Flax Darnel** *L. remotum*, a now rare weed of flax fields, is slenderer and usually unawned, with widely spaced spikelets △e. **4b* Stiff Darnel** *L. rigidum* has young leaves rolled together and glumes often completely hiding the spikelets △f, which are sunk in cavities in their common stalk, but lemmas not swollen in fruit.

4b-f

4a-e

50

1

2

3

4

The Grass Family Gramineae

Vulpia fescues are *annuals*, from 3–5 cm to 50–60 cm tall, with rather narrow leaves, flat when fresh. Inflorescence a usually *1-sided* panicle or raceme, the spikelets with very *unequal* glumes and both upper glume and lemma usually *long-awned*; flowering May–July. Dry bare places.

△ ×2

1.* **Dune Fescue** *Vulpia fasciculata* (incl. *V. membranacea*). Inflorescence like a miniature but 1-sided Wall Barley (p. 82), turning orange-brown in fruit, the lower part often *partly included in*, or scarcely separated from, the *inflated* leaf-sheath. Spikelets △a with a long upper glume and a *very short* lower glume; ovary downy at the tip. Sandy grassland, especially coastal dunes. B, F. Maps 16 and 17. Hybridises with Red Fescue (p. 48).

2.* **Squirreltail Fescue** *Vulpia bromoides*. Generally the commonest *Vulpia*, often on waste ground. Inflorescence *always* a panicle, well above the upper leaf-sheath; spikelets △b with the lower glume *at least half* the length of the upper. T. Map 18.

3.* **Ratstail Fescue** *Vulpia myuros*. Inflorescence often *drooping*, usually *partly included* in the *inflated* upper leaf-sheath; spikelets △c with the lower glume *less than half* as long as the upper, and the lemmas sometimes with marginal hairs (*V. megalura*). B, F, G. Map 19.

4.* **Bearded Fescue** *Vulpia ciliata*. Leaves often purplish. Inflorescence *reddening* as it ages, *well above* the upper leaf-sheath; spikelets △d with the lower glume minute, *much shorter* than the upper (but not as short as in Dune Fescue (1)). Often on coastal sand or shingle. B, F. Map 20.

5.* **Mat-grass Fescue** *Vulpia unilateralis* (*Nardurus maritimus*). Inflorescence somewhat like Mat-grass (p. 106), a short (5–20 cm) *tight raceme* well above the upper leaf-sheath, the spikelets △e *shorter-awned* than other *Vulpias* (the lemmas sometimes almost unawned) and the lower glume *more than half* as long as the upper. Chalk grassland, sometimes on railway tracks and on mine spoil tips in the south. B, F. Map 21.

6. Gravel Fescue *Micropyrum tenellum*. Differs from Mat-grass Fescue (5) especially in its 2-rowed spike. Spikelet △f, with the 2 glumes more or less *equal* or the lower slightly shorter, and the lemmas often *unawned*. Dry places, especially on sand, gravel or shingle. F, G. Map 22.

1-a

2-b

3-c

4-d

5-e

6-f

3 1 4 5 6

2

The Grass Family Gramineae

△ ×10

1-a

2-b

1: Fern Grass *Desmazeria rigida.* Annual, with *stiff* stems to 15 cm, long-persisting when dead. Leaves flat, often purplish. Inflorescence a *partial panicle,* looser, broader and with more spikelets than Sea Fern Grass (2), with which it may grow; spikelets with upper glume 2.3 mm or more, lower glume 2 mm or more and lemma △a *twice as long* as broad; flowering May–July. Dry, bare or sparsely grassy places, often on chalk or sand. B, F, G. Map 23.

2: Sea Fern Grass *Desmazeria marina.* Greener and fleshier than Fern Grass (1), with the tighter and *narrower* inflorescence usually a 2-sided *raceme;* spikelets with upper glume 2.3 mm or less, lower glume 2 mm or less and lemma △b *nearly as broad* as long; flowering May–July. Similar in habitat to Fern Grass, but always *by the sea.* Map 24.

3: Annual Meadow-grass *Poa annua.* The commonest and most widespread grass of bare and disturbed ground; *annual* or short-lived perennial, with stems slightly compressed, to 30 cm but usually much shorter, and occasionally rooting. Leaves *pale* green, often wrinkled. Inflorescence a *pyramidal* panicle, the lower branches more or less *down-turned* in fruit; spikelets along the *outer half* of each branch, whitish, sometimes tinged purple; flowering January–December. Probably derived from a hybrid between Creeping Meadow-grass and Early Meadow-grass (4). T. Map 25. **3a Creeping Meadow-grass** *P. supina* □ is stouter, creeping and always perennial, with the spikelets crowded at the tip of each branch; flowering March–May. Grassy, often damp places, especially in mountains. F, G, S. Map 26.

4: Early Meadow-grass *Poa infirma.* Differs from Annual Meadow-grass (3) in being always annual and having stems more strongly compressed, leaves narrower and *yellowish*-green, and narrower, less pyramidal panicles, their branches more or less *erect* in fruit and spikelets spaced out along the *whole length* of the branches; flowering March–May. Sandy places, usually *by the sea.* B, F. Map 27.

5: Flattened Meadow-grass *Poa compressa.* An often tufted perennial, with rather stiff, markedly *flattened* stems to 40 cm. Leaves flat, greyish. Inflorescence a rather *narrow* panicle, the spikelets along the whole length of the branches; flowering June–August. Dry, bare places, including walls, rocks and, less often, waste ground. T. Map 28.

6: Bulbous Meadow-grass *Poa bulbosa.* Tufted perennial; stems to 40 cm, *bulbous* at the base. Leaves greyish, sometimes folded. Inflorescence a *tight,* almost oval panicle, often purplish; spikelets along the whole length of the branches; flowering March–May. A viviparous form (see Viviparous Fescue, p. 48) is not uncommon. Dry, often sandy places, especially *by the sea,* also frequently on chalk or limestone. T. Map 29.

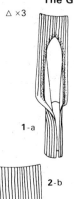

△ ×3

1-a

2-b

2-c

2a-d
△ ×1

3-e

4-f

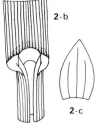

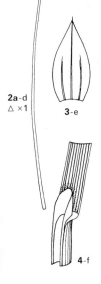

The Grass Family Gramineae

1. **Rough Meadow-grass** *Poa trivialis*. A common, loosely tufted perennial, with short *stolons*; stems to 90 cm, but usually prostrate at base. Leaves pale green, flat, 2–4.5 mm wide, rather soft, with sheaths slightly *rough*; ligule △a long, *pointed*. Inflorescence a narrowly pyramidal panicle; flowering May–July. Widespread in grassy places, often damp and shady and then producing few flowers. T. Map 30.

2. **Smooth Meadow-grass** *Poa pratensis*. A common, loosely tufted perennial, with long creeping *rhizomes*; stems to 50 cm. Leaves bright green, flat or grooved, 2–3 mm wide, stiffer than Rough Meadow-grass (1), with *smooth* sheaths; ligule △b a short *blunt* collar. Inflorescence a narrowly pyramidal panicle, the lower branches *3–5 together,* glumes bluntly pointed △c; flowering April–June. Grassy places, often dry and sunny, including wall tops; commoner in the south. T. Map 31. Known as Kentucky Blue-grass in North America. **2a* Narrow-leaved Meadow-grass** *P. angustifolia* is more tufted, has long, much narrower (*c.* 1 mm) wiry greyish leaves △d and grows in dry, especially lime-rich grassland. Map 32. **2b Northern Meadow-grass** *P. alpigena* has its leaves *abruptly* contracted to a *hooded apex* and the lower panicle branches usually *in pairs*; occasionally viviparous (see Viviparous Fescue, p. 48); flowering April–June. Hybridises with Alpine Meadow-grass (p. 58), when more often viviparous. Grassland and heathland. S. Map 33.

3. **Spreading Meadow-grass** *Poa subcaerulea*. Differs from Smooth Meadow-grass (2) in its shorter and usually *solitary* stems, greyish leaves with a *fringe of hairs* at the mouth of the sheath, and more sharply pointed glumes △e; flowering May–July. Grassland, often damp. T (distribution not fully known, but mainly in the north).

4. **Swamp Meadow-grass** *Poa palustris*. A loosely tufted perennial, not creeping; stems to 80 cm. Leaves *narrow*, 1–2 mm wide, with a *blunt* ligule △f, longer than Wood Meadow-grass (p. 58), and *smooth* sheaths. Inflorescence a narrowly pyramidal panicle, the spikelets well separated; lemmas bronze-tipped; flowering June–July. *Damp* grassy places, freshwater margins and woodland edges, occasionally on waste ground; easily overlooked. T. Map 34.

5. **Broad-leaved Meadow-grass** *Poa chaixii*. Much the *tallest* and broadest-leaved (6–10 mm wide) meadow-grass, a tufted perennial with stems to 120 cm and leaves abruptly contracted to a *hooded* tip. Inflorescence a loose, more or less pyramidal panicle with *straight* branches; spikelets with *unequal* glumes; flowering May–July. Woods, especially on mountains; also sown ornamentally, often with Various-leaved Fescue (p. 48). F, G, (B, S). Map 35. **5a** *P. remota* □ may have short stolons and has curved panicle branches, narrower leaves (to 8 mm), a longer ligule (2–3 mm) and more or less equal glumes. G, S. Map 36.

△ × 4

2-a

2a-b

3-c

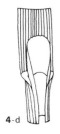

4-d

4a-e

1. **Wavy Meadow-grass** *Poa flexuosa.* Loosely tufted perennial; stems to 20 cm, with narrow (1–2 mm) leaves only towards their base, the top leaf sheathed. Inflorescence a compact oblong-ovoid panicle, its branches *grooved*, slightly *wavy* and each with only a few spikelets; flowering July–August. *Mountain* rocks and screes. B, S. Map 37. Hybridises with Alpine Meadow-grass (4). **1a** *P. laxa* □ is more tufted, with narrower leaves and a compacter inflorescence. F, G. Map 38.

2. **Glaucous Meadow-grass** *Poa glauca.* Tufted perennial, with stems (to 35 cm) and leaves (2.5–3 mm wide) both usually *glaucous*; ligules △a to 2 mm, *blunt.* Inflorescence a compact oblong-ovoid panicle, with few spikelets on each branch; flowering July–August. Occasionally (var. *balfourii*) stems and leaves green, when best distinguished from Wood Meadow-grass (3) by longer ligules and tighter inflorescence. Rocks and screes, usually in *mountains.* B, F, S. Map 39. **2a** *P. badensis* has green stems, stem-leaves folded and longer (2–6 mm) torn ligules △b. F, G. Map 40.

3. **Wood Meadow-grass** *Poa nemoralis.* A very variable, elegant, loosely tufted perennial, not creeping; stems to 70 cm. Leaves narrow, *c.* 2 mm, rather numerous, almost at right angles to the stem, with smooth sheaths and *very short* blunt ligules △c. Inflorescence oblong-ovoid, with numerous well-separated small spikelets, which are larger on mountains (cf. Glaucous Meadow-grass (2)); flowering May–July. *Woods* and shady places, sometimes on mountain and other rocks; in drier places a form with larger spikelets and slightly flattened stems occurs (cf. Flattened Meadow-grass, p. 54). T. Map 41.

4. **Alpine Meadow-grass** *Poa alpina.* Generally the commonest of the mountain meadow-grasses, a tufted perennial with stems to 30 cm. Leaves 3–4.5 mm wide, parallel-sided, the ligules △d 3–5 mm, often torn. Inflorescence a more or less *pyramidal* panicle, the branches mostly in pairs; spikelets purple-tinged; in Britain almost invariably viviparous (see Viviparous Fescue, p. 48), but flowering form commoner in Europe; flowering June–August. Grassland and rocky places on *mountains*: T. Map 42. **4a*** *P.×jemtlandica* (probably the hybrid of Alpine Meadow-grass with Wavy Meadow-grass (1)) is more tufted, with narrower leaves; narrowly triangular, shorter, pointed ligules △e; a looser, more or less ovoid inflorescence; and spikelets always viviparous. B, S. **4b Arctic Meadow-grass** *P. arctica* is shorter (to 15 cm), with narrower (2 mm) dark green leaves and unequal glumes. Dry grassland and heathland. S. Map 43.

The Grass Family Gramineae

SALTMARSH GRASSES *Puccinellia* are tufted perennials (except for Stiff Saltmarsh Grass (4)) and usually have erect to half-prostrate stems, greyish leaves and short blunt or rounded ligules. Inflorescence a panicle, the often purple-tinged spikelets with 3 or more florets; lemmas usually bluntly pointed; flowering June–July. Saline habitats. Various hybrids occasionally occur.

△ ×4

1-a

1-b
△ ×6

1a-c

1a-d
△ ×6

1. **Reflexed Saltmarsh Grass** *Puccinellia distans* ssp. *distans*. Stems to 65 cm; leaves flat; ligule △a. Inflorescence pyramidal, the branches *turned down* in fruit, their inner half *bare* of spikelets and *swollen* at the base; lemmas △b *rounded*. Drier saltmarshes, rocks and bare ground by the sea; increasingly by roadsides and around mineral workings inland. T. Map 44. **1a.** Northern Saltmarsh Grass *P.d.* ssp. *borealis* (*P. capillaris*) □ is shorter, with leaves often folded; longer, more pointed ligules △c; inflorescence narrow and compact, with shorter branches not down-turned in fruit; and bluntly pointed lemmas △d. B, G, S, South to Scotland and the Netherlands.

2. **Borrer's Saltmarsh Grass** *Puccinellia fasciculata*. Stems to 85 cm; leaves flat or folded. Inflorescence compact, *1-sided*, with branches *erect*, only the longer ones bare of spikelets at the base. Rather dry, muddy, often trampled places by the sea. B, F, G. Map 45. In *P.f.* ssp. *pseudodistans*, which prefers wetter places, the inflorescence is not one-sided and has spreading branches.

3. **Common Saltmarsh Grass** *Puccinellia maritima*. The only member of the genus to form a *sward*, its stolons *un-sheathed*. Stems to 30 cm; leaves flat or folded, sometimes dark green. Inflorescence narrow, and with spikelets *all along* the branches, which are *always erect*. Wetter saltmarshes, forming almost pure swards; sometimes inland. T. Map 46. **3a** *P. phryganodes*, with sheathed stolons and shorter stems, is a shy flowerer that grows in the far north. S. Map 47.

4. **Stiff Saltmarsh Grass** *Puccinellia rupestris*. Annual or biennial, with often *prostrate* stems to 40 cm; leaves flat. Inflorescence stiff, *1-sided*, with spikelets *all along* the short branches, somewhat recalling Fern Grass (p. 54); flowering May–August. Bare damp places by the sea. B, F, G. Map 48.

5. **Spiked Phippsia** *Phippsia algida*. Tufted perennial; stems to 8 cm, usually *no longer than* the narrow leaves, often half-prostrate. Inflorescence a spike-like panicle, the purplish spikelets each with *only 1 floret*; flowering July. Bare damp ground, often on mountains. S. Map 49. **5a** *P. concinna* □ has taller stems, well above the leaves, and pyramidal inflorescences. Map 50.

The Grass Family Gramineae

1. Arctic Marsh Grass *Arctophila fulva*. Stout perennial, with a thick rootstock; stems erect, to 80 cm. Leaves flat, *purplish,* the upper usually *longer* than the lower, those on non-flowering shoots alternate; ligule long, blunt, torn. Inflorescence a loose panicle, with long branches, the lower often *drooping;* both glumes and lemmas with *pale tips;* flowering July. Wet places in the far north. S. Map 51.

2.* Cocksfoot *Dactylis glomerata*. A very common, stout, tufted perennial; stems to 100 cm. Leaves flat, or with inrolled margins, often roughish; ligules long. Inflorescence a *1-sided* panicle, with the often purplish spikelets *densely clustered:* spikelets shortly awned; flowering late April—November. Grassy places, often sown. T. Map 52. **2a** var. *aschersoniana,* with looser panicles, fewer spikelets in each cluster and shorter awns, grows in woods in central Europe and is sown ornamentally elsewhere. Map 53.

3.* Crested Dogstail *Cynosurus cristatus*. A common tufted perennial, with wiry stems usually 50—60 cm, narrow leaves and short blunt ligules. Inflorescence a *compact 1-sided* spike-like panicle, the spikelets shortly awned, occasionally viviparous (see Viviparous Fescue, p. 48); flowering June—August. Meadows and pastures on a wide range of soils. T. Map 54.

4.* Rough Dogstail *Cynosurus echinatus*. *Annual,* with stems to 60 cm and rather broad leaves. Inflorescence a *1-sided* spike-like panicle, the *long-awned* spikelets in dense *ovoid* or even globose *heads;* flowering June—July. Dry bare places, including sandy fields and waste ground. (B, F, G). Map 55.

1

4

2

3

The Grass Family Gramineae

△ ×6

1a-a

3-b

1.* Whorl Grass *Catabrosa aquatica.* *Creeping* peren-nial; stems to 70 cm. Leaves *broad*, soft, to 10 mm wide, blunt-tipped, sweet-tasting; ligules long, blunt. Inflorescence an oblong panicle, the branches in *alternate half-whorls*; spikelets with *2 florets*, *unawned*, purplish; flowering May–July. *Wet places*, mud, especially in or by shallow fresh water. T. Map 56. **1a*** *C.a.* ssp. *minor* is a distinctive sprawling plant with shorter stems, leaves and inflorescences, the spikelets △a 1-flowered. Coastal sand flushed by fresh water in north and west Britain.

2. Broad-leaved Cinna *Cinna latifolia.* Loosely tufted perennial; stems *tall*, to 130 cm. Leaves *very broad*, to 20 mm, roughish; ligules long, torn. Inflorescence a loose panicle, the spikelets with *1 floret*, *awned*; flowering July. Damp woods. S. Map 57.

3.* Loose Silky-bent *Apera spica-venti.* Annual, with stems to 100 cm. Leaves variable, to 10 mm wide, flat, roughish, the sheaths roughish and often purplish; ligules bluntly pointed. Inflorescence a *pyramidal* panicle, with spreading branches, the spikelets △b each with *1 floret* and a *very long awn*, to 12 mm; flowering June–July. Sandy fields and waysides. T, but (S). Map 58.

4.* Dense Silky-bent *Apera interrupta.* Annual, with stems to 70 cm. Leaves rather narrow, to 4 mm, flat (but inrolled when dry), roughish, with sheaths smooth and often purplish; ligules pointed. Inflorescence a *narrow* panicle, the spaces between the branches giving it an interrupted appearance; the spikelets with *1 floret* and a *very long awn*, to 10 mm; flowering June–July. Bare, sandy ground. T, but (B). Map 59.

5.* Early Sand-grass *Mibora minima.* One of the smallest European grasses; annual, with thread-like stems to 15 cm, but often only *2–3 cm*. Leaves mostly in a basal tuft, short, slender and often slightly greyish. Inflorescence a *1-sided spike-like* raceme, the spikelets with 1 floret; flowering *February*–May. Bare, sandy ground, often by the sea. B (rare), F, G. Map 60.

The Grass Family Gramineae

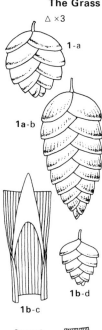

△ ×3

1-a

1a-b

1b-c

1b-d

1: **Quaking Grass** *Briza media.* Loosely tufted perennial with *non-flowering* shoots; stems usually 20–50 cm. Ligules short, blunt. Inflorescence a pyramidal panicle, with very distinctive *ovoid to broadly triangular* spikelets △a, 4–12 mm, usually purplish, shaking in the wind on slender stalks; flowering May–August. Grassland, especially *calcareous*. T. Map 61. **1a* Great Quaking Grass** *B. maxima* is annual, with all shoots flowering, nodding panicles and much larger spikelets △b, 14–25 mm, often red-brown or silvery green. Bare ground; grown for ornament and so an occasional casual. (B, F) north to north England. **1b* Lesser Quaking Grass** *B. minor* is annual, with all shoots flowering, a longer ligule △c (to 6 mm) and smaller (3–5 mm) spikelets △d usually green. Bare and disturbed ground. B, F. Map 62.

2: **Blue Moor Grass** *Sesleria albicans.* Loosely tufted perennial; stems to 45 cm. Leaves green or *greyish*. Inflorescence a rather tight *ovoid* spike-like panicle strongly tinged *bluish-purple*, but occasionally greenish-white; flowering April–June. Dry calcareous grassland, in Britain only on limestone. B. F. G. Map 63. **2a** *S. caerulea* is extremely similar, but has greyer leaves and a looser inflorescence, and prefers damper grassland in the lowlands. G, S. Map 64.

3: **Mountain Melick** *Melica nutans.* Creeping perennial; stems to 60 cm. Ligule △e very short, blunt. Inflorescence a 1-sided *nodding raceme* or partial panicle; lemmas matt; flowering May–July. Open woods and shady rocks on limestone. T. Map 65. **3a** *M. picta* is extremely similar, but has a longer, more oval ligule △f and shining lemmas. Dry open woods. G. Map 66.

4: **Wood Melick** *Melica uniflora.* Creeping perennial; stems to 60 cm. Leaves rather pale green, their sheaths △g with a short *bristle* on the side opposite the blade. Inflorescence a very loose panicle, the spikelets *sparsely* distributed along the branches; flowering May–July. Dry woods, especially beechwoods, and shady banks. T. Map 67.

5. **Ciliate Melick** *Melica ciliata.* Tufted or creeping perennial; stems to 60 cm. Leaves 1–4 mm wide, stiff, grooved, the midrib *obscure*. Inflorescence a loose, often *nodding* panicle; lemma △h with a fringe of hairs; flowering June–July. Dry, rocky and waste places on limestone; occasional on walls. F, G, S. Map 68. **5a** *M. transsilvanica* □ has broader (2–6 mm) leaves, usually flat and with a prominent midrib, a tighter erect panicle and the upper lemmas usually longer than the lower. F, G. Map 69.

3-e

3a-f

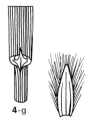

4-g

5-h

5

5a

1

3

4

2

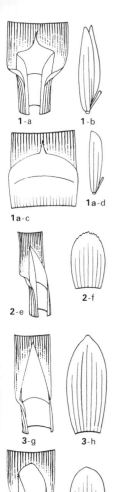

1-a 1-b

1a-d

1a-c

2-e

2-f

3-g

3-h

4-i

4-j

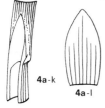

4a-k 4a-l

5-m 5-n

The Grass Family Gramineae

1.* Reed Sweet-grass *Glyceria maxima.* One of the 3 *tall* waterside grasses (with Common Reed and Reed Canary-grass, p. 102); a stout *patch-forming* perennial, with stems to 200 cm; leaves *broad*, to 18 mm, bright green, with a *brown mark* at the stem-junction; ligules △a short, rounded, with a *sharp point*. Inflorescence a well-branched panicle, the spikelets unawned with *3* stamens; floret △b; flowering June–August. Freshwater margins and wet marshy areas. T. Map 70. **1a Northern Sweet-grass** *G. lithuanica* □ is smaller (to 150 cm), with leaves to 9 mm wide, ligule △c, inflorescence a nodding panicle with numerous hair-like branches, and floret △d with 2 stamens. Wet *woods*. S. Map 71.

2.* Small Sweet-grass *Glyceria declinata.* Loosely tufted perennial; stems often *curved*, to 50 cm. Leaves glaucous, *abruptly* pointed at tip; ligules △e pointed, *shorter* than Floating Sweet-grass (3). Inflorescence a narrow, almost spike-like panicle, the lemmas △f *distinctly toothed*; flowering June–August. Muddy freshwater margins, shallow ponds. T. Map 72.

3.* Floating Sweet-grass *Glyceria fluitans.* Generally the commonest of the smaller sweet-grasses; a spreading perennial with rather weak stems, to 120 cm. Leaves green or glaucous, often *floating* on the water surface; ligules △g long, *acutely* pointed. Inflorescence a narrow, nodding panicle, with elongated spikelets, the lemmas △h *pointed*; flowering late May to August. Hybridises with Plicate Sweet-grass (4) to yield Hybrid Sweet-grass (4a). In and by still and slow-flowing fresh water. T. Map 73.

4.* Plicate Sweet-grass *Glyceria plicata.* Loosely tufted perennial; stems to 80 cm. Leaves green or glaucous; ligules △i short, *bluntly* pointed. Inflorescence a more branched and *spreading* panicle than Small and Floating Sweet-grasses (2 and 3), the lemmas △j *blunt*; flowering June–August. Freshwater margins and other wet places, avoiding poor, acid waters. T. Map 74. Hybridises with Floating Sweet-grass to give **4a.* Hybrid Sweet-grass** *G.× pedicillata*, which is intermediate in many ways between its parents, especially in its more branched and spreading panicle than Floating Sweet-grass, its oblong ligules △k and its bluntly pointed lemmas △l. Like other hybrids it has sterile anthers, i.e. lacking pollen, and no proper fruits. Spreads vegetatively and occurs in the absence of both parents.

5. Swamp Grass *Scolochloa festucacea.* Tall creeping perennial; stems to 200 cm. Differs from Reed Sweet-grass (1) in its narrower (to 12 mm), *rough-edged* leaves, longer (to 6 mm) ligules △m and *no brown marks* on stem; lemma △n. Still and slow-moving fresh water. S. Map 75.

Ligules △ ×2
Florets, lemmas △ ×6

The Grass Family Gramineae

1. **Great Brome** *Bromus diandrus.* Annual, with stems to 90 cm, *hairy at the top.* Leaves and sheaths hairy. Inflorescence a loose *nodding* panicle, the long branches mostly with only 1 *very long* (50–70 mm) spikelet, the awns *very long*, 35–65 mm; flowering May–July. Roadsides and bare, often sandy ground. (B, F). Map 76.

△ ×1

2. **Stiff Brome** *Bromus rigidus.* Annual, with stems to 40 cm, *downy at the top.* Leaves and sheaths hairy. Inflorescence an *erect* panicle, with branches *shorter* than the 25–35 mm spikelets, and awns *long*, 30–50 mm; flowering May–July. Bare and waste ground. (B), F. Map 77.

3. **Barren Brome** *Bromus sterilis.* Much the commonest of the long-awned annual bromes; stems *hairless*, to 50 or 100 cm. Leaves soft, hairy, with downy sheaths. Inflorescence a *nodding* panicle, with usually *only 1* spikelet △a (20–35 mm) on each fairly long branch, and awns 15–30 mm; flowering late April to July. Roadsides, field margins and other bare or waste ground. T. Map 78.

4. **Drooping Brome** *Bromus tectorum.* Annual, with stems *hairless* or almost so, to 90 cm. Leaves and sheaths hairy. Inflorescence a panicle, becoming *nodding*, with *up to 4* spikelets (10–20 mm), which are occasionally hairy, on each fairly long branch, the awns 10–18 mm; flowering May–July. Waysides, bare and waste ground. T, but (B). Map 79.

3-a

5. **Compact Brome** *Bromus madritensis.* Annual, with stems usually *hairless*, to 60 cm. Leaves hairless or downy, the lower sheaths hairy, the upper often hairless. Inflorescence a more or less *dense* panicle, with *very short* branches bearing *1–2* erect spikelets, each 30–50 mm long, the awns 12–20 mm; flowering May–July. Dry, rocky banks, old walls, bare and waste places. (B), F. Map 80.

The Grass Family Grameneae

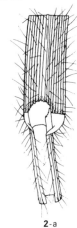

△ ×3

1. **Hungarian Brome** *Bromus inermis.* Perennial with a creeping rootstock; stems to 150 cm. Leaves to 9 mm wide; leaves, sheaths and stems usually all *hairless.* Inflorescence a spreading panicle, the spikelets usually *unawned*, the awn, if present, short; anthers large, yellow; flowering June—July. Bare, often sandy ground; spreading north and west. T, but (B, S).

2. **Hairy Brome** *Bromus ramosus.* A common grass of *shady places*; a loosely tufted perennial with stems to 190 cm. Leaves lax, to 15 mm wide, somewhat hairy, the sheaths covered with *long white hairs*, which distinguish it from Giant Fescue (p. 46); ligule △a to 6 mm. Inflorescence a well-branched, very *gracefully drooping* panicle, the lowest node with only *2* branches and a *hairy* scale at its base; spikelets often purplish, *long-awned*; flowering July—August, persisting dried through the winter. Woods, shady hedge banks. T. Map 81.

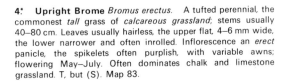

2-a

3. **Lesser Hairy Brome** *Bromus benekenii.* A tufted perennial, intermediate in appearance between Hairy Brome (2) and False Brome (5); stems to 120 cm. Leaves to 12 mm wide, sparsely hairy; sheaths shortly hairy, but the *topmost* often *hairless*; ligules △b to 3 mm. Inflorescence much *less spreading* and drooping than Hairy Brome, with *3–5* branches at the lowest node, which has a *hairless* scale at its base; spikelets *long-awned*; flowering July—August. Woods, especially of beech, mainly on chalk and limestone. T. Map 82.

4. **Upright Brome** *Bromus erectus.* A tufted perennial, the commonest *tall* grass of *calcareous grassland*; stems usually 40—80 cm. Leaves usually hairless, the upper flat, 4—6 mm wide, the lower narrower and often inrolled. Inflorescence an *erect* panicle, the spikelets often purplish, with variable awns; flowering May—July. Often dominates chalk and limestone grassland. T, but (S). Map 83.

3-b

5. **False Brome** *Brachypodium sylvaticum.* A very common grass of *shady places*; a tufted perennial with lax stems to 100 cm. Leaves *lax*, often hairy, to 12 mm wide, with sheaths usually hairy. Inflorescence a *nodding raceme* with awned and stalked spikelets; flowering July—August. Confusable with Bearded Couch (p. 78), which has hairless leaves and unstalked spikelets. Woods, scrub, shady hedge banks, often persisting in the open when scrub is cleared, when the leaves may be yellowish-green. T. Map 84.

6. **Tor Grass** *Brachypodium pinnatum.* *Patch-forming* perennial; stems stiff, to 120 cm. Leaves *stiff*, 3—6 mm wide, flat or inrolled, often conspicuously *yellowish-green* (cf. False Brome (5) growing in the open), the sheaths usually hairless. Inflorescence an *erect raceme*, the spikelets shortly awned; flowering June—August. *Calcareous* grassland. T. Map 85.

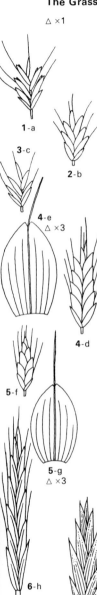

△ ×1

1-a

3-c

2-b

4-e
△ ×3

4-d

5-f

5-g
△ ×3

6-h

6a-i

1.* **Field Brome** *Bromus arvensis.* Annual, with stems to 110 cm. Leaves and sheaths sparsely hairy, or the upper sheaths hairless. Inflorescence an *open* panicle, erect at first but drooping later, the long (10–25 mm) narrow, often purplish spikelets △a on *long, slender* branches, the lemmas *c. 6 mm* long, awned, widely *divergent* when mature; flowering June–August. Cultivated, waste and other bare ground. T. Map 86.

2.* **Rye Brome** *Bromus secalinus.* A rather variable annual, with stems usually 50–100 cm. Leaf-sheaths usually *hairless* or the lower sparsely hairy. Inflorescence a loose, *1-sided* and often nodding panicle, with spikelets △b 10–30 mm, hairy or not; lemmas *7–9 mm* long, awned or not, widely *divergent* when mature; flowering July–August. Mainly a cornfield weed, also on waste ground. (T). Map 87. **2a Great Rye Brome** *B. grossus* has very large spikelets (18–50 mm), with glumes as well as lemmas awned and longer (10–14 mm) lemmas. Belgium only, in spelt (p. 80) crops; almost extinct. Note: *B. bromoideus*, another spelt weed of Belgium, with lemmas appearing 3-awned, is extinct in the wild.

3.* **False Rye Brome** *Bromus pseudosecalinus.* Annual, intermediate in appearance between Rye and Smooth Bromes (2 and 5); stems rather *short*, to 60 cm. Leaf-sheaths *all hairy*. Inflorescence a *narrow* panicle, *erect* at first but nodding later, with *short* (8–12 mm) *hairless* awned spikelets △c and lemmas 5–6 mm long and widely *divergent* when mature; flowering June–August. Grassy places, especially by tracks and roads, at present only known from Britain and Ireland.

4.* **Meadow Brome** *Bromus commutatus.* Annual, with stems usually 40–80 cm. Leaf-sheaths hairy, the *upper* usually *hairless.* Inflorescence a *loose* panicle, erect at first but nodding later, the spikelets △d 15–25 mm long, hairy or not, with the mature lemmas △e 8–11 mm long, *angled* and *not* diverging; flowering May–July. Intermediates with Smooth Brome (5) occur. Grassy places, especially moist meadows, also on road verges and in waste places. T, but (S). Map 88.

5.* **Smooth Brome** *Bromus racemosus.* Has a *narrower* and more erect panicle than Meadow Brome (4), with spikelets △f *smaller* (10–15 mm) and *always hairless*, the *ovate* lemmas △g *c.* 7 mm and *hooded* at the tip; flowering June–July. Habitat as Meadow Brome, with which intermediates occur. T. Map 89.

6.* **California Brome** *Bromus carinatus.* Loosely tufted perennial; stems to 80 cm. Leaf-sheaths *hairless.* Inflorescence a loose nodding panicle, with hairless spikelets △h 25–30 mm long, lemmas *c.* 16 mm and *awns* 4–10 mm; flowering May onwards. Waysides and waste ground; naturalised in Britain and the Netherlands only. **6a* Rescue Brome** *B. willdenowii* (*B. unioloides*) may be taller, with the lower leaf-sheaths hairy, the panicle looser, the spikelets △i longer (to 40 mm) and broader, and the awn absent or very short. (B, F).

The Grass Family Gramineae

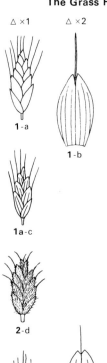

△ ×1   △ ×2

1-a

1-b

1-a-c

2-d

3-e

3-f

4-g

4-h

5-i

5-j

1. **Soft Brome** *Bromus hordeaceus* ssp. *hordeaceus* (*B. mollis*). Generally the commonest of the shorter-awned annual bromes; very variable, with stems to 100 cm, but often much shorter and can be only 5 cm. Leaves with hairy sheaths. Inflorescence an erect, usually fairly tight panicle, with spikelets △a *12–25 mm* long, usually *hairy*, and *bluntly* angled lemmas △b 8–11 mm long with a *narrow* pale margin, the awns *straight*, 4–11 mm long; flowering May–July. Grassy places, especially older swards in hayfields and on road verges, also dry rocky ground. T. Map 90. **1a.** **Lesser Soft Brome** *B. h.* ssp. *thominii* is much *shorter* (1–8 cm) and differs also in having shorter (8–18 mm) and narrower spikelets △c, sometimes *hairless*, with shorter (*c.* 7 mm) lemmas and *divergent* awns 3–7 mm long. Intermediates often occur. *Sand dunes* by the sea. B, F. Map 91. **1b.** **Hybrid Soft Brome** *B. × pseudothominii* is intermediate in most of its dimensions between its parents, Soft and Slender Soft Bromes (1 and 3), and may have hairy or hairless spikelets with the pale lemma-margin usually narrow. Grassy places, especially newer swards such as leys and new road verges. Note: *Interrupted Brome *B. interruptus*, with a narrow, interrupted panicle, formerly found in sainfoin fields in England, appears to be extinct in the wild.

2. **Least Soft Brome** *Bromus hordeaceus* ssp. *ferronii*. Much *shorter* than typical Soft Brome (1) (2–15 cm); differs also in having *densely* hairy spikelets △d 8–18 mm long, lemmas 7–8 mm long and *curved* awns 3–5 mm long; flowering May–July. *Cliff tops* by the sea. B, F. Map 92.

3. **Slender Soft Brome** *Bromus lepidus*. Differs from Soft Brome (1) in having spikelets △e usually *hairless* (but beware hairless forms of Soft Brome), shorter (5–15 mm) and narrower; the shorter (*c.* 6 mm) lemmas △f *sharply* angled and with a *broad* pale margin; and shorter (2–5 mm) awns. Hybridises with Soft Brome to give Hybrid Soft Brome (1b). Grassy places, especially newer swards, such as leys and new road verges. (T). Map 93.

4. **Japanese Brome** *Bromus japonicus*. Annual, with stems to 80 cm. Leaves and sheaths hairy. Inflorescence a spreading panicle, nodding, with *long branches* and spikelets △g 20–40 mm, hairy or not, the *lanceolate* lemmas △h up to 4 mm wide, usually longer than the *divergent* awns; flowering May–June. Grassy places, often on sand. (B), F, G. Map 94.

5. **Rough Brome** *Bromus squarrosus*. Annual, with stems to 60 cm. Leaves and sheaths hairy, often long-haired. Inflorescence a *1-sided raceme* or loose partial panicle with shortish branches, *flattened* spikelets △i up to *70 mm* long and *broad* (to 7 mm) *angled* lemmas △j, longer than the awns; flowering May–June. Grassy places, often on sand or gravel. (B), F. Map 95.

3 4 1 2 5

The Grass Family Gramineae

1. **Lyme Grass** *Leymus arenarius* (*Elymus arenarius*). Stout *bluish-grey* perennial; stems to 150 cm. Leaves *broad*, to 15 mm, the margins inrolled. Inflorescence a spike, the spikelets in *pairs* alternating up opposite sides of the axis; flowering June–August. *Sand* dunes and sandy shores, mainly *by the sea*; sometimes planted as a stabiliser. T. Map 96.

COUCHES *Elymus* were formerly *Agropyron*. Inflorescence a spike, the spikelets *solitary*, alternating up opposite sides of the axis.

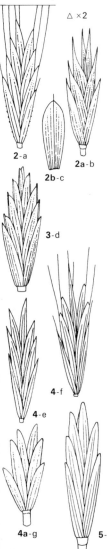

△ ×2

2. **Bearded Couch** *Elymus caninus* (incl. *Agropyron donianum*). A variable, loosely tufted perennial; stems to 110 cm, usually *downy* at the leaf-junctions. Leaves flat. Spikelet △a, glumes with a *narrow* pale margin, shortly awned, the lemmas with a *long awn* to 18 mm; flowering June–August. Woods and other, often damp, shady places; rocky places in hills. T. Map 97. **2a** *E. mutabilis* is an Arctic species, usually hairless at the leaf-junctions, spikelet △b with much shorter awns. S. Map 98. **2b** *E. alaskanus* is also Arctic; its glumes △c have a broad pale margin and its awns are much shorter; more open, rocky or stony places. S. Map 99.

2-a

2a-b

2b-c

3. **Sea Couch** *Elymus pycnanthus* (*Agropyron pungens*). A stiff, patch-forming, *glaucous* perennial; stems to 120 cm. Leaves *sharply* pointed, up to 6 mm wide, usually *tightly* inrolled, with prominent *crowded* flattened veins. Spikelet △d, with glumes *pointed*, lemmas sometimes awned to 10 mm; flowering June–August. Hybridises with Common and Sand Couches (4 and 5). Sand, shingle and mud, *by the sea*. B, F, G. Map 100. **3a** *E. pungens* (*A. campestre*) has broader leaves, their veins further apart, its glumes blunter and its lemmas never awned. F. Map 101.

3-d

4. **Common Couch** *Elymus repens* (*Agropyron repens*). A very common and tenacious creeping perennial weed, also known as Twitch; stems to 120 cm. Leaves usually *green* and *flat*, sparsely white-hairy above. Spikelet △e and f, with glumes more or less *pointed*, sometimes shortly awned, the lemmas usually *unawned*; flowering June–August. Hybridises with Sea and Sand Couches (3 and 5), and with Meadow Barley (p. 82). Widespread on waste and cultivated ground. T. Map 102. **4a** **Hairy Couch** *E. hispidus* has leaves usually glaucous and inrolled, spikelet △g with both glumes and lemmas blunt. Dry, usually sandy or stony places. F, G.

4-f

4-e

5. **Sand Couch** *Elymus farctus* (*Agropyron junceiforme*). A *glaucous* patch-forming perennial; stems to 60 cm. Leaves to 6 mm wide, flat or with the hardened margins inrolled. Spikelets △h *hardly overlapping* each other up the axis, which *breaks easily* just above each one; glumes and lemmas *blunt*; flowering June–August. Hybridises with Sea and Common Couches (3 and 4). *Sand by the sea*, to seaward of Sea Couch. T. Map 103.

4a-g

5-h

The Grass Family Gramineae

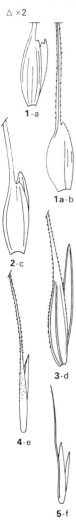

△ ×2

1. **Bread Wheat** *Triticum aestivum.* A *commonly* culti-vated annual; stems *hollow,* to 150 cm. Leaves flat, to 16 mm wide. Inflorescence a spike, similar to the Couches (p. 78) but much *stouter,* the *solitary* spikelets alternating up the *hairless* axis; glumes *broad, blunt,* florets △a with the lemmas *broad,* usually *awnless,* but sometimes with an awn up to 16 cm; flowering June–August. A relic of cultivation by field and road margins and on waste ground. (T). **1a.** **Rivet Wheat** *T. turgidum* has solid stems, the spike axis hairy, the glumes keeled through-out and ending in a curved tooth, and the florets △b with lemmas always awned. Much less often cultivated. (B, F, G).

2. **Spelt Wheat** *Triticum spelta.* Annual, still cultivated in the hills in parts of north-west Europe. Differs from Bread Wheat (1) in having a slenderer spike that readily *falls apart* below each spikelet when mature; glumes *keeled throughout* and remaining attached to the seed, and florets △c with lemmas always *awned,* to 10 cm; flowering June–August. Another relic of cultivation, now rare and only on thin soils. (F, G).

3. **Rye** *Secale cereale.* An annual, less often cultivated than formerly. Differs from Bread Wheat (1) in having somewhat narrower leaves, to 10 mm, a *slenderer* spike, with the axis slightly *downy,* glumes *narrow, pointed* and keeled *throughout,* and florets △d with lemmas *narrow* and always with a shorter awn, to 5 cm. Like Bread and Spelt Wheats a relic of cultivation, usually on sandy soils. (T).

4. **Wood Barley** *Hordelymus europaeus.* Tufted perennial; stems to 110 cm, *downy* at the leaf-junctions. Leaves to 14 mm wide, with *hairy* sheaths, or the upper sometimes hairless. In-florescence a spike with clumps of 3 1-flowered spikelets alter-nating up the axis; glumes *narrow, awned,* the florets △e with lemmas *narrow* with an awn to 2.5 cm; flowering June–July. *Woods* and shady banks, especially beechwoods and on cal-careous soils. T. Map 104.

5. **French Oat** *Gaudinia fragilis.* Annual or biennial; stems to 120 cm. Leaves flat, downy, with downy sheaths; ligules very short. Inflorescence a long (to 35 cm) spike, the *solitary* spike-lets alternating up the axis, which readily *falls apart* above each one when mature; upper glumes blunt, florets △f with lower glumes pointed and lemmas with a short *bent awn;* flowering May–July. Grassy, often damp places. (B, F).

1-a

1a-b

2-c

3-d

4-e

5-f

1   2   3   5   4

The Grass Family Gramineae

BARLEYS *Hordeum* have leaves flat and the inflorescence a loose spike, the 1-flowered spikelets being arranged in *triplets*, alternating up the axis; glumes and lemmas *long-awned*.

△ ×2

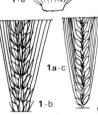

1.* **Six-rowed Barley** *Hordeum vulgare*. The commonest barley in cultivation; annual, with stems to 130 cm. Spike up to 12 cm, excluding *very long* awns, to 18 cm; spikelets △a *all fertile*, in *4 or 6* longitudinal rows △b; flowering June–August. A relic of cultivation, by field and road margins and on waste ground. (T). **1a.*** **Two-rowed Barley** *H. distichon* is shorter, with the spikelets △c in 2 rows, shorter awns and lemmas of outer spikelets blunt. A similar relic of cultivation.

2.* **Wall Barley** *Hordeum murinum* (incl. *H. leporinum*). The commonest of the wild barleys; annual, with stems to 50 cm. Leaves to 8 mm wide, with long, overlapping auricles. Spike to 7 cm long, excluding the awns (to 30 mm), the central un-stalked spikelet *only* of each triplet △d being *fertile*, the outer short-stalked ones either barren or male; the axis breaking readily above each triplet when mature; glumes △e of inner spikelet and inner glume of outer spikelets *lanceolate* and *fringed with hairs*, the outer glumes very narrow; flowering May onwards. Waste ground, waysides, often at the foot of walls. T, but (S). Map 105.

3.* **Sea Barley** *Hordeum marinum*. Differs from Wall Barley (2) in its generally narrower *glaucous* leaves, to 5 mm wide, lack of auricles, usually shorter spike, to 5 cm excluding awns, and having all glumes △f *narrow* and *hairless*, the inner ones of the outer spikelets *winged* at the base; flowering June–July. Bare and waste ground and grassy places *by the sea*. B, F, G. Map 106.

4.* **Meadow Barley** *Hordeum secalinum*. Tufted perennial; stems slender, to 70 cm. Leaves to 6 mm wide, sometimes greyish, with *short* auricles. Spike rather *short*, to 5 cm, excluding awns to 14 mm; central unstalked spikelet of each triplet △g fertile, the 2 outer short-stalked ones male or barren; glumes of outer spikelets *bristle-like*, lemmas lanceolate; flowering June–July. Hybridises with Common Couch (p. 78). *Established* grassland, meadows and pastures, on heavy soils. T. Map 107.

5.* **Foxtail Barley** *Hordeum jubatum*. Tufted perennial; stems to 60 cm. Leaves narrow, to 4 mm. Easily identified at a distance by the squirrel-tail appearance of its often nodding inflorescence, due to the *long whitish silky awns*; spikelets with glumes very narrow; flowering June–August. Widely naturalised from North America, and spreading, on waysides and in waste places. (T).

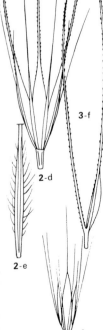

1-a

1a-c

1-b

3-f

2-d

2-e

4-g

5-h

1

3

2

4

5

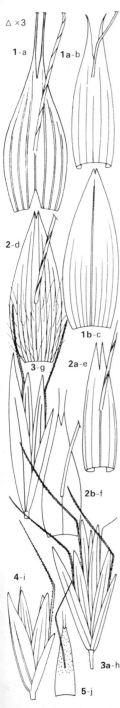

△ ×3

The Grass Family Gramineae

1. **Bristle Oat** *Avena strigosa*. Still cultivated in a few areas; annual with stems to 100 cm. Ligules bluntly pointed. Inflorescence a panicle, not spreading widely; spikelets 14–22 mm long with 2–3 florets, the lemmas △a *narrow*, hairy or not, *short toothed*, with 2 tiny bristles and a *bent* awn; flowering July–August. A relic of cultivation, sometimes mixed with Cultivated Oat (2a), and on waste ground. (T). **1a** *A. brevis* has shorter spikelets (10–12 mm) and almost hairless lemmas △b with 2 distinct teeth ending in tiny bristles. F, G, S. **1b** *A. nuda* has stems with a waxy bloom, longer (20–30 mm) spikelets and lemmas △c with 2 teeth but no bristles. (G). *A. brevis* and *A. nuda* occur rarely as relics, in NW and C Europe respectively.

2. **Wild Oat** *Avena fatua*. The commonest oat growing wild in most areas; annual with stems to 100 cm. Ligules rounded. Inflorescence a spreading panicle, the axis of the 2–3-flowered 18–25 mm spikelets readily *falling apart* when mature, both above the glumes and *between the florets*; lemmas △d *broad*, tipped with 2 small teeth, and with a basal tuft of *tawny hairs* and a long *bent* awn; flowering June–August. Arable fields and waste ground, often growing with barley and wheat. (T). **2a* Cultivated Oat** *A. sativa* lacks the tawny hairs and has the lemma △e tip sometimes only notched and the awn (on the lowest lemma only) usually straight. Hybridises with Wild Oat. Still widely cultivated, but a less frequent relic than formerly. **2b* Winter Wild Oat** *A. sterilis* (incl. *A. ludoviciana*) has narrower lemmas △f with a longer awn (to 60 mm) and the spikelet axis falling apart only above the glumes. Often on heavier soils. (B, F).

3. **Downy Oat-grass** *Avenula pubescens*. Loosely tufted perennial; stems to 100 cm. Leaves *green*, hairy, with *hairy* sheaths; ligules pointed. Inflorescence a panicle with 2–4-flowered silvery spikelets △g, the lemmas with a *bent* awn; flowering May–July. *Calcareous* grassland. T. Map 108. **3a* Meadow Oat-grass** *A. pratensis* has glaucous leaves, the leaves and sheaths both hairless, the panicle more spike-like, and the spikelets △h 4–6-flowered. Also in other dry grassland. Map 109.

4. **False Oat-grass** *Arrhenatherum elatius*. A very common, loosely tufted perennial; stems to 150 cm, shining, sometimes swollen and *bulbous* at the base, and with yellowish roots. Leaves usually hairless; ligules blunt. Inflorescence a spreading panicle, with shining *2-flowered* spikelets △i, the lemmas often 2-toothed at the tip, lower with a **straight** awn, upper usually awnless; flowering May–September. Grassland, waysides, waste ground. T. Map 110.

5. **Soft Bearded Oat-grass** *Ventenata dubia*. Annual; stems to 70 cm, *downy* at the leaf-junctions. Leaves and sheaths hairy or not; leaves grooved or folded; ligules pointed, to 9 mm. Inflorescence a loose panicle with *long* branches, the short, *sparse* (10–15 mm) spikelets 2–3-flowered; lemmas △j split, the lowest with a short bristle, the top one with 2 bristles and a *bent* awn; flowering June–July. Dry, rather sparsely grassy places. F, G. Map 111.

The Grass Family Gramineae

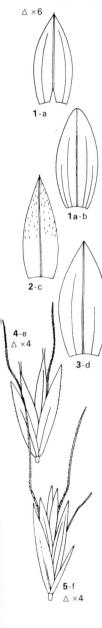

△ ×6

1-a

1a-b

2-c

3-d

4-e
△ ×4

5-f
△ ×4

1. **Crested Hair-grass** *Koeleria macrantha.* Loosely tufted perennial; stems to 40 cm. Leaves of non-flowering shoots *green* or glaucous, *flat* or *folded*, sometimes slightly rough, the old sheaths persistent, loosely overlapping. Inflorescence a spike-like panicle, the spikelets 2–5 mm long, green, brown or purplish, with a silvery margin, the glumes △a *pointed*, ending in a short bristle; flowering June–July. Dry, especially *calcareous* grassland. T, but (S). Map 112. Distribution not yet distinguishable from **1a*** **Glaucous Hair-grass** *K. glauca*, which has stems thickened at base, leaves of non-flowering shoots always glaucous, inrolled and rough to the touch, the old sheaths breaking down into filaments, the panicle interrupted below and the glumes △b blunt. Sandy soils, especially dunes. T. Map 112.

2. **Pyramidal Hair-grass** *Koeleria pyramidata.* Loosely tufted perennial; stems to 90 cm. Leaves of non-flowering shoots *flat*, green or somewhat glaucous, sometimes sparsely hairy. Inflorescence a *pyramidal* panicle, with spikelets 6–8 mm long, the lemmas △c sharply pointed or ending in a short bristle; flowering June–July. Dry, often rocky and calcareous, grassy places and wood margins. F, G, (S). Map 113.

3. **Somerset Hair-grass** *Koeleria vallesiana.* Tufted perennial, woody and *thickened* at the base, the overlapping old leaf-sheaths breaking down into a *dense basal network of fibres.* Leaves grey-green, either inrolled and *hair-like* or flat and broader. Inflorescence a tight spike-like panicle, the spikelets 4–6 mm, the glumes △d sharply pointed to shortly bristled, pale brown or shining green, with a *broad* pale margin; flowering June–August. Dry, rocky, calcareous grassland. B (rare), F. Map 114.

4. **Northern Oat-grass** *Trisetum spicatum.* Tufted perennial; stems to 40 cm, *downy* at the top. Inflorescence a tight spike-like panicle, sometimes interrupted; spikelets △e; lemmas with a markedly *curved* awn; flowering July–September. Rocky and grassy places, in mountains or on tundra. F, G, S. Map 115. **4a** *T. subalpestre* is more loosely tufted with hairless stems but downy leaf-sheaths and a looser panicle with pale shining glumes. By running fresh water. Map 116.

5. **Yellow Oat-grass** *Trisetum flavescens.* A common, loosely tufted perennial; stems to 80 cm. Inflorescence a loose but not widely spreading panicle, the numerous *yellowish* 2–4-flowered spikelets △f with a conspicuous *bent* awn. Flowering June–July. Grassland, especially calcareous. T, but (S). Map 117.

6. **Harestail Grass** *Lagurus ovatus.* Annual, with stems to 50 cm. Leaf-sheaths inflated. Inflorescence a very distinctive dense *egg-shaped* panicle, *whitish, soft and woolly*, like a rabbit's scut, the lemmas with long white awns to 20 mm; flowering June–August. Mainly on sand *by the sea*, also on waste ground inland. (B), F, mainly southern. Map 118.

2

6

4

5

1

3

The Grass Family Gramineae

△ ×6

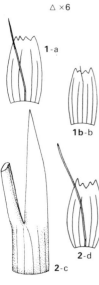

1-a

1b-b

2-d
2-c

3-e·

3-f

4-g

1. **Tufted Hair-grass** *Deschampsia cespitosa.* A common, tall, *conspicuously tufted* perennial; stems to 200 cm. Leaves usually channelled, *rough-edged* and darker green than Tall Fescue (p. 46); ligules long, pointed. Inflorescence a spreading panicle, the spikelets silvery or purplish, the lemmas △a *blunt, equally* toothed at the tip, with a *straight* awn; flowering June—August. Widespread in damp woods and grassland, marshes. T. Map 119. **1a* Alpine Hair-grass** ssp. *alpina* □ has most spikelets viviparous (see Viviparous Fescue, p. 48), a less spreading panicle and pointed lemmas. Bare stony places on mountains. B, S. **1b** *D. media* is more loosely tufted, with shorter stems (to 70 cm), ligules and awns, and the outer teeth of the lemma △b longer than the inner. F, G. Map 120.

2. **Bog Hair-grass** *Deschampsia setacea.* Densely tufted perennial; stems to 70 cm. Leaves hair-like, roughish, with long *pointed* ligules △c. Inflorescence a spreading panicle, the spikelets purple and yellowish; lemmas △d blunt, with the outer teeth *longer* than the inner, and a long *bent* awn; flowering July—August. *Bogs*, peaty pools on wet heaths; becoming rare. T. Map 121.

3. **Wavy Hair-grass** *Deschampsia flexuosa.* Tufted perennial; stems to 100 cm. Leaves *hair-like*, with short *blunt* ligules △e. Inflorescence a gracefully spreading panicle, the purplish or silvery spikelets; lemmas △f blunt, *minutely toothed* at the tip, with a long *bent* awn; flowering May—July. *Dry* heaths, woods and moorland on *acid* soils. T. Map 122.

4. **Arctic Hair-grass** *Vahlodea atropurpurea.* Loosely tufted perennial; stems *short*, to 30 cm. Leaves flat, to 3 mm wide; ligules short. Inflorescence a loose panicle, the lower branches often *down-turned*, the glumes purple with a broad brownish margin; lemmas △g blunt, *untoothed, hairy* and very shortly awned; flowering July. Damp grassy places on *acid* soils. S. Map 123.

5. **Early Hair-grass** *Aira praecox.* A short annual, most conspicuous before flowering, when the inflorescences are concealed in *silvery sheaths*; stems to 12 cm. Leaves channelled, with *smooth* sheaths. Inflorescence a *spike-like* panicle, the lemmas shortly awned; flowering *April*—June. Dry bare ground, especially on *acid sandy* soils. T. Map 124.

6. **Silver Hair-grass** *Aira caryophyllea.* Annual, with stems to *50 cm*. Leaves channelled, with *roughish* sheaths. Inflorescence a *spreading* panicle, the lemmas shortly awned; flowering May—July. Ssp. *caryophyllea* has stems 5-35 cm and a tighter inflorescence with spikelets 2.5-3.5 mm and their longer stalks usually more than 5 mm. Ssp. *multiculmis* has stems usually 20—50 cm and a looser inflorescence with spikelets 2—2.5 mm and their longer stalks usually less than 5 mm. Dry bare places on sandy and gravelly soils. T. Map 125.

The Grass Family Gramineae

 △ ×6

2-a

3-b

1. Southern Holy Grass *Hierochloë australis.* Aromatic *tufted* perennial; stems to 60 cm. Leaves flat, often *bluish.* Inflorescence a pyramidal panicle, the lemmas with a *short* bent awn, to 3 mm, those of the male florets cleft at the tip; flowering *April—May. Woods.* G. Map 126. **1a Arctic Holy Grass** *H. alpina* is shorter (to 35 cm), with channelled leaves and longer awns. Mountains and tundra. S. Map 127.

2.* Holy Grass *Hierochloë odorata.* Aromatic creeping perennial; stems to 60 cm. Leaves flat, bright *green,* dull beneath. Inflorescence a pyramidal panicle, the spikelets rounded, the lemmas sometimes with a *very short* bent awn, to 1 mm, those of the male florets cleft at the tip △a; flowering late *March* to June. Riversides, lake margins and *wet* peaty *grassland.* T, but B (rare). Map 128.

3.* Sweet Vernal Grass *Anthoxanthum odoratum.* A common, variable, tufted perennial, one of the earliest grasses to flower, aromatic when dried and smelling of new-mown hay; stems to 50 cm. Leaves flat, with non-flowering shoots present at flowering. Inflorescence a *spike-like* panicle, the lemmas △b with a *bent* awn; flowering April—July. Widespread in grassland. T. Map 129. **3a* Annual Vernal Grass** *A. aristatum (A. puelii)* is annual, with all shoots flowering, and looser inflorescences. A weed of bare, usually sandy ground. (B, F, G). Map 130.

4.* Yorkshire Fog *Holcus lanatus.* A very common, greyish, *softly downy* tufted perennial; stems to 100 cm, *not* bearded at joints. Inflorescence a spreading panicle, with whitish or pinkish-purple spikelets, the lemmas △c with a *hooked* awn; flowering May—August. Grassland and waste ground, often forming extensive stands. T. Map 131.

5.* Creeping Soft-grass *Holcus mollis.* Slenderer and greener than Yorkshire Fog (4); differs also in having a *creeping* rootstock, conspicuously *bearded* stem joints and *bent* awns (lemma △d); flowering June—August. Open woods, shady hedge banks, heaths, bare sandy ground, preferring *acid* soils. T. Map 132.

6.* Grey Hair-grass *Corynephorus canescens.* Markedly *tufted* perennial; stems to 60 cm. Leaves *glaucous, hair-like,* with purplish-pink sheaths. Inflorescence a contracted panicle, the spikelets 2-flowered and purplish, the lemmas △e with an awn *thickened* at the tip; flowering June—July. Acid sandy soils, in Britain rare and almost exclusively on coastal dunes. T. Map 133.

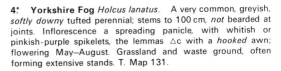

4-c

5-d

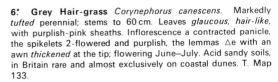

6-e

The Grass Family Gramineae

△ ×3

BENTS *Agrostis* are usually perennials, with leaves flat and panicles with *1-flowered* spikelets at the end of branches arranged in whorls.

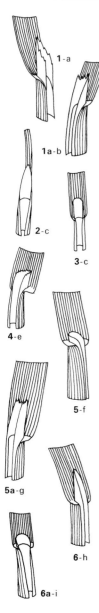

1-a

1a-b

2-c

3-c

4-e

5-f

5a-g

6-h

6a-i

1.* Velvet Bent *Agrostis canina.* Loosely tufted, with *short* leafy *runners*; stems to 70 cm. Leaves narrow, 1–2 mm wide; ligules △a long, *pointed*. Panicle loose, becoming *contracted* in fruit, spikelets purple-brown, lemmas *awned*; flowering June–August. *Damp* grassy places. T. Map 134. **1a* Brown Bent** *A. vineale* (*A. montana*) is more tufted with no runners, ligules △b sometimes blunt and a denser panicle. Drier places, especially on heaths and moors, mainly in the north and west.

2.* Bristle Bent *Agrostis curtisii* (*A. setacea*). Conspicuously tufted, with stems to 60 cm. Leaves *hair-like*, often greyish; ligules △c *pointed*. Panicle *contracted*, spikelets yellow-green, lemmas *awned*; flowering June–July. Dry heaths and moors. B, F. Map 135.

3. Arctic Bent *Agrostis mertensii.* Tufted, with stems to 30 cm. Leaves narrow, 1–3 mm wide; ligules △d *blunt*. Panicle *spreading* in fruit, spikelets dark purplish-brown, lemmas *awned*; flowering July. Tundra. S. Map 136.

4. Northern Bent *Agrostis clavata.* Usually *annual*, with stems to 70 cm. Leaves to 4 mm wide; ligule △e. Panicle *spreading*, lemmas *unawned*; flowering July. Disturbed ground. S. Map 137. **4a* Ticklegrass** *A. scabra*, a perennial with awned lemmas, is an increasing casual. (B, G).

5.* Common Bent *Agrostis capillaris* (*A. tenuis*). A common perennial with a creeping rootstock; stems to 70 cm. Leaves to 4 mm wide; ligules △f *short, blunt*. Panicles *always spreading*, spikelets greenish- or purplish-brown, lemmas usually unawned; flowering June–August. Widespread in grassland, especially on acid soils. T. Map 138. **5a* Black Bent** *A. gigantea* is much taller, to 150 cm, with broader leaves, to 6 mm and ligules △g of non-flowering shoots usually longer than wide. Waysides, waste and cultivated ground; a common weed of cereal crops. Map 139.

6.* Creeping Bent *Agrostis stolonifera.* A very common *creeping* perennial, the runners *long*, up to 200 cm; stems to 100 cm. Ligules △h *pointed*, 2–7 mm long. Panicles spreading in flower, *contracted* in fruit, spikelets greenish to purplish, lemmas usually *unawned*; flowering June–August. Hybridises with Annual Beard-grass (see Perennial Beard-grass, p. 94). Grassland, dry to wet, waysides and disturbed ground. T. Map 140. **6a** *A. castellana* has shorter runners and ligules △i, and awned lemmas. An increasing casual. (F).

The Grass Family Gramineae

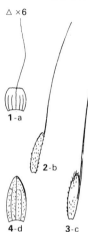

△ ×6

1-a

2-b

4-d

3-c

5-e

6-f

6a-g

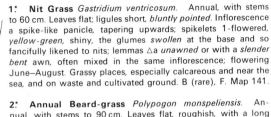

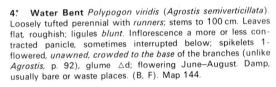

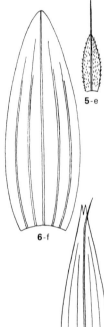

1. **Nit Grass** *Gastridium ventricosum.* Annual, with stems to 60 cm. Leaves flat; ligules short, *bluntly pointed.* Inflorescence a spike-like panicle, tapering upwards; spikelets 1-flowered, *yellow-green*, shiny, the glumes *swollen* at the base and so fancifully likened to nits; lemmas △a *unawned* or with a *slender bent* awn, often mixed in the same inflorescence; flowering June–August. Grassy places, especially calcareous and near the sea, and on waste and cultivated ground. B (rare), F. Map 141.

2. **Annual Beard-grass** *Polypogon monspeliensis.* Annual, with stems to 90 cm. Leaves flat, roughish, with a long *pointed* ligule. Inflorescence a dense spike-like panicle, to 16 cm long, the spikelets 1-flowered, greenish; glumes △b *notched* at the tip and with a *long* silky awn, to 7 mm, making the inflorescence look bearded; lemmas shortly awned; flowering June–August. Bare or sparsely grassy places, especially *by the sea.* B (rare), F. Map 142.

3. **Southern Beard-grass** *Polypogon maritimus.* Differs from Annual Beard-grass (2) in its *shorter* stems (to 25 cm) and often purplish panicles (to 5 cm), the glumes △c *cleft* at the tip and the lemmas *unawned*; flowering June–August. Bare or sparsely grassy places by the sea. F. Map 143.

4. **Water Bent** *Polypogon viridis* (*Agrostis semiverticillata*). Loosely tufted perennial with *runners*; stems to 100 cm. Leaves flat, roughish; ligules *blunt.* Inflorescence a more or less contracted panicle, sometimes interrupted below; spikelets 1-flowered, *unawned, crowded to the base* of the branches (unlike *Agrostis*, p. 92), glume △d; flowering June–August. Damp, usually bare or waste places. (B, F). Map 144.

5. **Perennial Beard-grass** × *Agropogon littoralis.* The hybrid between Annual Beard-grass (2) and Creeping Bent (p. 92) is perennial, with stems to 50 cm. Leaves roughish, often greyish; ligules *bluntly* pointed. Inflorescence a contracted, usually more or less lobed panicle, the glumes △e *not notched* and with a *short* awn, the lemmas *awned*; flowering June–September. Saltmarshes and other wet, bare places *by the sea.* B (rare), F. Map 145.

6. **Marram** *Ammophila arenaria.* The grass which binds younger coastal sand dunes together with its long tough rhizomes; a coarse perennial with stems to 120 cm. Leaves greyish-green, to 5 mm wide, *sharply* pointed, the margins *inrolled.* Inflorescence a stout, fairly tight spike-like panicle, with 1-flowered straw-coloured spikelets and *unawned* lemmas △f; flowering June–August. *Dunes*, often planted. T. Map 146.
6a **Hybrid Marram** × *Ammocalamagrostis baltica* □ the hybrid between Marram and Wood Small-reed (p. 96), differs from Marram in its flat leaves, more open purplish inflorescences and awned lemmas △g, and from Wood Small-reed in its narrower leaves, less open panicle and habitat. Coastal sand dunes; a sand binder but much less often planted than Marram. Map 147.

94

The Grass Family Gramineae

SMALL-REEDS *Calamagrostis* are *tall* tufted perennials with flat or channelled, usually *hairless* (except Narrow Small-reed (3)) leaves and ligules usually *blunt*. Inflorescence a panicle, the 1-flowered spikelets usually *purple* or purplish-brown, with glumes more or less *equal* and usually *lanceolate* and the *awned* lemmas usually *cleft* at the tip and with a *tuft of hairs* at their base.

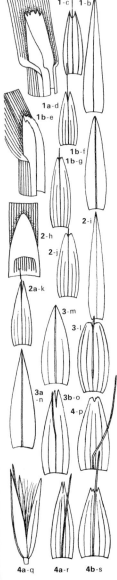

1.* Wood Small-reed *Calamagrostis epigeios*. Stems to 200 cm, with 2–4 joints. Leaves more or less flat, 4–20 mm wide; ligules △a *torn*. Inflorescence *erect*, fairly open, the spikelets sometimes greenish or silvery-grey; glumes △b *narrowly* lanceolate, lemmas △c *3-veined*, with a *straight* awn, *shorter* than the basal hairs; flowering June–August. Hybridises with Purple, Narrow and Rough Small-reeds (2, 3 and 4) and with Marram (p. 94). Damp woods, fens, ditches, usually on heavy soils. T. Map 148. **1a** *C. pseudophragmites* has stems with 2–3 joints, leaves to 10 mm wide, the panicle nodding, unequal glumes and a longer awn; lemma △d. F, G. Map 149. **1b** *C. villosa* has stems with 4–5 joints, leaves to 10 mm wide, torn ligules △e, nodding panicles, broader glumes △f and 5-veined lemmas △g. G. Map 150.

2.* Purple Small-reed *Calamagrostis canescens*. Stems to 150 cm, with *4–6* joints, often *branched*. Leaves to 8 mm wide, shortly *hairy* above; ligule △h. Inflorescence open, glumes △i narrowly lanceolate, lemmas △j *5-veined* with a *very short* straight awn and *notched* at the tip, equalling the basal hairs; flowering June–July. Hybridises with Wood and Rough Small-reeds (1 and 4). Fens, marshes, wet woods. T. Map 151. **2a** *C. purpurea* has stems with 6–8 joints, leaves hairless and to 8 mm wide, a longer awn and basal hairs sometimes longer than lemma △k. F, G, S. Map 152.

3.* Narrow Small-reed *Calamagrostis stricta*. Stems to 100 cm, with 2–3 joints. Leaves to 5 mm wide. Inflorescence *contracted*, the spikelets 3–4 mm long, lemmas △l with a *straight* awn, *longer* than their basal hairs, glumes △m broad lanceolate; flowering June–August. Hybridises with Wood Small-reed (1). Bogs, fens, marshes. B, G, S. Map 153. **3a* Scottish Small-reed** *C. scotica* has longer (5–6 mm) spikelets and narrower, more pointed glumes △n. Scotland only. Map 154. **3b** *C. lapponica* has longer (4–8 mm) spikelets and blunt lemmas △o with a bent awn and equalling the basal hairs. Tundra, dry heaths and woods. S. Map 155.

4. Rough Small-reed *Calamagrostis arundinacea*. Stems to 150 cm, with 2–3 joints. Leaves flat, to 12 mm wide, *hairy* at base. Inflorescence open or contracted, the lemmas △p *longer* than the basal hairs, the awns *bent*; flowering June–August. Hybridises with Wood and Purple Small-reeds (1 and 2). Could be confused with Tall Fescue (p. 46), which has hairless leaves. Deciduous woods, marshes. F, G, S. Map 156. **4a** *C. varia* has leaves rarely hairy at base, spikelets △q yellowish-brown, and lemmas △r with a somewhat shorter awn and longer basal hairs. Also on mountains, often on calcareous soils. F, G, S (rare). Map 157. **4b** *C. chalybaea* has stems usually with 4 joints, torn ligules and lemmas △s with longer awns and basal hairs. S. Map 158.

Ligules △ ×3
Glumes, lemmas, spikelet △ ×6

1

3

2

4

The Grass Family Gramineae

CATSTAILS and **TIMOTHY** *Phleum* are tufted perennials (except Sand Catstail (4)), with flat leaves. Inflorescence a *dense spike-like* panicle, with 1-flowered spikelets, shortly *awned* glumes and *unawned* lemmas. Foxtails *Alopecurus* (p. 100) have similar inflorescences, but the glumes are unawned and the lemmas shortly awned.

△ ×6

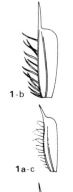

1ᵕ Timothy *Phleum pratense.* A common fodder plant, the tallest *Phleum*, with stems *to 150 cm.* Ligules △a *blunt.* Inflorescence *cylindrical,* to 15 cm long (or in cultivars even longer), sometimes tinged purple, the glumes △b blunt with a 1–2 mm awn; flowering June–August. Grassland, waysides, waste ground; frequently sown. T. Map 159. **1aᵕ Smaller Catstail** *P.p.* ssp. *bertolonii,* with shorter stems (to 70 cm), panicles (to 8 cm) and awns (glume △c), and pointed ligules △d, grows in natural grassland, especially on calcareous soils.

2ᵕ Alpine Catstail *Phleum alpinum.* Stems to 50 cm. Leaf-sheaths somewhat inflated; ligules short, blunt. Inflorescence more or less *ovoid,* to 5 cm long, usually purplish, with blunt glumes △e and awns to 4 mm; flowering July–August. (Alpine Foxtail (p. 100) has pointed glumes and inconspicuous or no awns.) Wet grassy places, wet flushes and wet rock ledges in *mountains.* T. Map 160.

3ᵕ Purple-stem Catstail *Phleum phleoides.* Stems to 90 cm, often *purple.* Ligules short, blunt. Inflorescence to 10 cm long or rarely more, *slender, tapering* towards the tip, often purplish; glumes △f *abruptly* narrowed to a *very short* awn; flowering June–August. Dry, especially calcareous, sandy grassland. T, but B (rare). Map 161.

4ᵕ Sand Catstail *Phleum arenarium. Annual,* with no non-flowering shoots at flowering time; stems to 35 cm. Leaf-sheaths inflated; ligules △g *bluntly pointed,* to 7 mm long. Inflorescence *short,* 1–5 cm, tapering at each end, sometimes purplish; glumes △h *gradually* narrowing to a short awn; flowering May–June, turning straw-coloured later. Sand and shingle *by the sea,* rarely on sand inland. T. Map 162.

1

2

3

4

The Grass Family Gramineae

△ ×6   **FOXTAILS** *Alopecurus* have flat leaves and usually blunt ligules. Inflorescence a *dense spike-like* panicle with 1-flowered spikelets, differing from the catstails *Phleum* (p. 98) in having awns on the lemmas but not on the glumes.

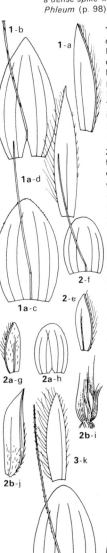

1: Meadow Foxtail *Alopecurus pratensis.* A common, *early-flowering* loosely tufted perennial; stems to 110 cm Leaves 4–8 mm wide; ligules *short, c.* 2 mm. Inflorescence cylindrical, to 9 cm long (longer in cultivars), with glumes △a and lemmas △b both *pointed,* a conspicuous awn and orange or purplish anthers; flowering April–June. Hybridises with Marsh Foxtail (2). Grassland, especially on neutral soils. T. Map 163. **1a Reed Foxtail** *A. arundinaceus* has broader leaves (to 12 mm), inflated leaf-sheaths, a longer ligule (to 5 mm), blunt lemmas △c with inconspicuous awns; glume △d; flowering May–July. Damper, often saline grassland. F (rare), G, S. Map 164.

2: Marsh Foxtail *Alopecurus geniculatus.* Creeping or floating perennial, the stems *'kneed'* and *rooting* at the base and often also bent at the joints higher up, to 45 cm. Inflorescence cylindrical, to 6 cm long, the glumes △e and lemmas △f both *blunt,* awns *inconspicuous* on spike and anthers *yellow* or purplish; flowering June–August. Hybridises with Meadow Foxtail (1). *Wet* grassy places and shallow still water. T. Map 165. **2a* Orange Foxtail** *A. aequajis*☐ is often annual, with more pointed ligules, inconspicuous awns, paler and more shining glumes △g (lemma △h) and orange anthers. Especially on drying mud. Map 166. **2b* Bulbous Foxtail** *A. bulbosus* is tufted and erect, with stems △i bulbous at the base and not rooting at the joints, and the glumes △j pointed. Saltmarshes and grassy places by the sea. B, F, G. Map 167. Note: Meadow and Marsh Foxtails occasionally have bulbous stems.

3: Alpine Foxtail *Alopecurus alpinus.* Loosely tufted perennial; stems usually 10–30 cm. Inflorescence *ovoid,* with *pointed* glumes △k and blunt lemmas △l, *unawned* or with an inconspicuous awn (Alpine Catstail (p. 98) has blunt glumes and long awns); flowering June–August. Wet grassy places and rocks on *mountains.* B. Map 168. **3a** *A. rendlei* is annual, to 20 cm, and has inflated leaf-sheaths and glumes △m abruptly contracted above the middle. Damp meadows and roadsides on calcareous soils. F, G. Map 169.

4: Black Grass *Alopecurus myosuroides.* A notorious arable weed, also called Black Twitch; annual; with stems to 85 cm. Inflorescence *tapering* towards tip, to 10 cm long, with pointed glumes △n, blunt lemmas △o and awns *conspicuous;* flowering May–August. Cultivated and other *disturbed* ground. T, but (S). Map 170.

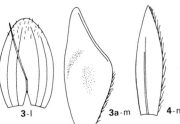

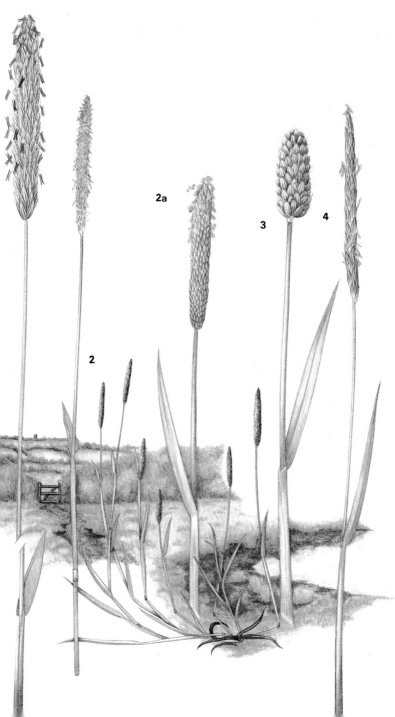

2a

2

3

4

The Grass Family Gramineae

△ ×6

4-a

4a-b

6-c

1. **Hard Grass** *Parapholis strigosa*. Annual, with *stiff* straight stems to 40 cm; hard to detect when not in flower. Ligules very short. Inflorescence a *straight* spike, the 1-flowered spikelets *sunk into their axis*; anthers *2 mm*; flowering June–August. Bare and sparsely grassy places in saltmarshes and elsewhere *by the sea*. B, F, G. Map 171.

2. **Curved Hard-grass** *Parapholis incurva*. Differs from Hard Grass (1) in being usually shorter and having the topmost leaf-sheath *inflated*, the spikes *curved* and the anthers only *1 mm*; flowering June–August. Similar in habitat to Hard Grass. B, F. Map 172.

3. **Reed Canary-grass** *Phalaris arundinacea*. A *tall* stout creeping perennial; stems to 200 cm. Leaves *rough-edged*, to 18 mm wide, occasionally (especially in garden forms) variegated with yellow, and persisting through the winter when dead; ligules *long*, torn. Inflorescence a compact *lobed* panicle, the spikelets *2–3-flowered*, pale purple, white or pale yellow, *unawned*; flowering June–August. Marshes, fens, damp woods, water meadows, freshwater margins. T. Map 173.

4. **Canary Grass** *Phalaris canariensis*. Annual, with stems to 120 cm, but usually much shorter. Leaves to 10 mm wide, the upper sheaths inflated; ligules pointed. Inflorescence a tight *ovoid* panicle, the spikelets 2–3-flowered and *7–9 mm* long; glumes △a with an *untoothed* wing, anthers *3–4 mm*; flowering June–September. A common bird-seed relic, often on rubbish dumps and in gardens where birds are fed. (B, F, G). **4a.** **Lesser Canary-grass** *P. minor* often has more elongated inflorescences, with shorter spikelets (*c.* 5 mm), glumes △b with a toothed wing, and shorter anthers (*c.* 1 mm). Casual on dry bare ground. **4b.** **Bulbous Canary-grass** *P. aquatica* is an often taller perennial, with a swollen stem-base, a cylindrical panicle and tiny or no lower lemmas. Uncommon but increasing casual. (B).

5. **Sheath Grass** *Coleanthus subtilis*. Annual, with *threadlike* stems to 8 cm. Leaves narrow, usually *recurved*, the sheaths *inflated*, the topmost *infolding* the lower part of the inflorescence, which is a raceme of oval or rounded clusters of 1-flowered spikelets, the lemmas *awned*; flowering June–July. Damp mud by still fresh water. F (rare), G (rare). Map 174.

6. **Common Reed** *Phragmites australis*. A common, *very tall*, stout, coarse perennial, forming *extensive beds* with its creeping rootstock; stems to 350 cm or more, persisting through the winter as hard canes. Leaves greyish, smooth-edged to 5 *cm* wide, collapsing in winter; ligule a *line of hairs*. Inflorescence a spreading panicle, the numerous unawned, usually dark purple spikelets △c with 2–10 florets and *long* silky hairs along the axis; flowering August–October. Fens, marshes, swamps, freshwater or brackish margins, damp clayey slopes; sometimes persisting in apparently unsuitable habitats, new shoots arising from the extensive rhizomes. T. Map 175.

The Grass Family Gramineae

1.⁎ **Wood Millet** *Milium effusum.* *Tall,* loosely tufted perennial; stems to 180 cm. Leaves *pale green,* flat, to 15 mm wide; ligules bluntly pointed. Inflorescence a spreading panicle, with well-spaced, pale green, 1-flowered, *unawned* spikelets; flowering May–July. *Woods,* especially on damp and calcareous soils. T. Map 176.

2. **Early Millet** *Milium vernale.* *Annual;* stems, usually more or less *prostrate,* to 50 cm, but often much shorter. Leaves narrow, with sheaths often purplish and short, pointed ligules. Inflorescence a spreading panicle, with *unawned,* 1-flowered spikelets; flowering *April*–May. *Sandy* places, especially on dunes. F, G. Map 177.

FEATHER-GRASSES *Stipa* are perennials, with leaves folded or channelled, usually with hairless outer and downy inner surfaces. Inflorescence a contracted panicle with 1-flowered spikelets; glumes with a long narrow tip, and lemmas with *immensely long* bent *awns,* the lower part usually hairless and the upper usually conspicuously *feathery* with pinnately arranged hairs. Dry, stony, steppe-like grassland.

3. **Golden Feather-grass** *Stipa pulcherrima.* Stems to 100 cm. Leaves with shining *yellowish* sheaths, which are fringed with hairs; ligules short on non-flowering stems, to 7 mm on flowering stems. Spikelet △a; awns to 50 cm, their hairs to *c. 7 mm* long; flowering May–June. F, G. Map 178. **3a** *S. bavarica,* very local in south Germany, has the outer leaf surface hairy, the sheaths downy and the awn hairs 3–4 mm. Map 179.

4. **Bristle-leaved Feather-grass** *Stipa tirsa.* A fairly tall, stout plant. Leaves thread-like with the outer surface *downy* and a *long bristle-like* tip, the lower to 100 cm long; sheaths hairless, ligules very short. Spikelet △b; awns to *c.* 40 cm, the hairs 5–6 mm; flowering May–July. F, G. Map 180. **4a** *S. joannis* has much less attenuated, greyish leaves, with the outer surface hairless and the tips hairy. F, G, S. Map 181.

5. **Hair-like Feather-grass** *Stipa capillata.* Stems to 70 cm. Ligules of lower leaves very short, of upper *long,* to 20 mm. Spikelet △c; awns *hairless,* to 18 cm; flowering July–August. F, G. Map 182.

6. **Rough Feather-grass** *Achnatherum calamagrostis.* Stout tufted perennial; stems to 120 cm, with numerous *scale-like sheaths* at the base. Leaves channelled, gradually narrowed to a *long slender* tip; ligules very short. Inflorescence a panicle, with shining, often purplish 1-flowered spikelets △d, *hairy* lemmas and a straight or curved awn to *10 mm;* flowering June–September. Rocky or stony places, in calcareous soils, especially in mountains. F, G. Map 183.

The Grass Family Gramineae

△ ×6

1. **Pampas Grass** *Cortaderia selloana*.   A large, stout, coarse perennial, forming the substantial *tussocks* familiar in gardens; stems to 300 cm. Leaves glaucous, with edges *sharp* enough to cut human skin. Inflorescence a spreading panicle, *silvery white* with long silky hairs, the glumes long-awned; flowering August–October in Europe. Occasionally naturalised. (B, F).

2. **Heath Grass** *Danthonia decumbens* (*Sieglingia decumbens*).   Tufted perennial, the tufts often *flattened*; stems to 60 cm, often half-prostrate. Leaves pale green, usually flat; ligule a *ring of hairs*. Inflorescence a raceme or few-branched panicle, the almost globular spikelets 4–6-flowered, *unawned*; flowering June–August. Heaths, *acid* grassland, especially on sand or peat. T. Map 184.

5-a

3. **Purple Moor-grass** *Molinia caerulea*.   Tussock-forming perennial; stems to 90 cm. Leaves flat, greyish, *3–6 mm* wide, not persisting through the winter; ligule a *ring of hairs*. Inflorescence a spike-like panicle, the 1–4-flowered spikelets usually *purple* but sometimes green or yellowish; anthers *purple-brown*; flowering July–September. *Wet* heaths and moors on acid soils with a fluctuating water table. T. Map 185. **3a** *M.c.* ssp. *arundinacea* is taller, to 250 cm, with leaves 8–12 mm wide and spreading panicles. Marshes and fens on less acid soils.

4. **Mat Grass** *Nardus stricta*.   Tufted perennial; stems to 20 cm. Leaves *wiry*, grey-green; ligules short, blunt. Inflorescence a *one-sided* spike, the spikelets 1-flowered, in 2 rows, the lemmas shortly *awned*; flowering June–August. Heaths, moors, mountains, in very *infertile*, often peaty or sandy soils; the commonest plant species in many upland areas. T. Map 186.

5. **Hairy Love-grass** *Eragrostis pilosa*.   Annual; stems hairless, to 50 cm. Leaves flat, with *no glands*; ligule a *ring of hairs*. Inflorescence a spreading panicle, the lower branches *hairy*, the spikelets △a 5–10-flowered, unawned, the glumes *unequal*; flowering July–October. Disturbed, often sandy or stony ground. F, G. Map 187.

6-b

6. **Greater Love-grass** *Eragrostis cilianensis*.   Annual; stems sometimes hairy, to 60 cm. Leaves flat, with *raised glands* on margins and midrib; ligule a *ring of hairs*. Inflorescence a rather contracted *hairless* panicle, the spikelets △b up to *40-flowered*, the glumes more or less *equal*, the lemmas *2 mm or more*, unawned; flowering May–September. Cultivated and disturbed ground, often on sandy soils. (B, F, G). **6a Lesser Love-grass** *E. minor* is smaller, with stems always hairy, a looser panicle, the spikelets △c 5–12-flowered and lemmas 2 mm or less; flowering July–September. Sometimes on road verges and railway tracks. (F, G). Map 188. Many other *Eragrostis* species are rare casuals.

6a-c

The Grass Family Gramineae

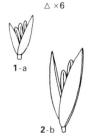

△ ×6

1-a

2-b

1. False Foxtail *Crypsis alopecuroides.* Annual; stems to 30 cm, *bent* at leaf-junctions. Leaves green or slightly glaucous. Inflorescence a purplish *spike-like* panicle, its base *covered* by the topmost leaf-sheath; spikelets unawned, the florets △a with *3 stamens*; flowering August–September. Damp places, wet in winter. F. Map 189.

2. Sharp-leaved Grass *Crypsis aculeatus.* Annual; stems to 20 cm, half-prostrate. Leaves glaucous, *sharply pointed.* Inflorescence a series of flattened or *globular* green *heads* up the stem, *enfolded* by the 2–3 topmost leaf-sheaths; spikelets unawned, the florets △b with *2 stamens*; flowering July–August. Damp places, usually saline. F. Map 190.

BRISTLE-GRASSES *Setaria* are annuals with stems to over 100 cm, but often much shorter, and flat leaves, variable in width. Inflorescence a *spike-like* panicle, the spikelets with 2 florets, only the upper fertile; conspicuous *bristles* (usually 3–10 mm long) under each spikelet persist on the axis after the spikelet falls; flowering July–October. Cultivated and waste ground, often on rubbish dumps, commonest in the south.

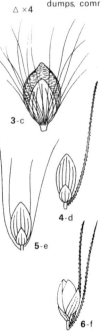

△ ×4

3-c

4-d

5-e

6-f

3. **Yellow Bristle-grass** *Setaria pumila* (*S. glauca*; *S. lutescens*). Stems to 130 cm; leaves 2–10 mm wide. Spikelets △c with *4–12 yellowish* bristles, the upper lemma *longer* than the upper glume. (B, F, G).

4. **Rough Bristle-grass** *Setaria verticillata.* Leaves 4–16 mm wide. Spikelets △d often clustered up the panicle axis, appearing *whorled*; bristles usually *only 1* per spikelet; upper lemma *equalling* upper glume. Hybridises with Green Bristle-grass (5). An increasing casual. (B, F, G).

5. **Green Bristle-grass** *Setaria viridis.* Leaves 4–12 mm wide. Inflorescence sometimes purplish, the spikelets △e with *1–3* bristles, the upper lemma *equalling* the upper glume. Hybridises with Rough Bristle-grass (4). (B), F, G, (S).

6. **Foxtail Bristle-grass** *Setaria italica.* Known as Italian Millet in bird-seed mixtures. Stems to 150 cm; leaves sometimes very *broad*, 6–20 mm. Panicle broad, often yellowish, lobed, often *nodding* at tip, commonly sold intact for feeding to canaries; bristles very variable, *2–16 mm* long, 2–5 per cluster of spikelets △f; upper lemmas equalling or longer than upper glumes; seed *detaches* from rest of spikelet when ripe. Especially in or near gardens of aviculturists. (B, F, G).

The Grass Family Gramineae

△ ×6

1-a

2-b

1. **Bermuda Grass** *Cynodon dactylon*. Far-*creeping* perennial; stems to 30 cm. Leaves flat; ligule a *ring of hairs*. Inflorescence a fan of finger-like *spikes*, clustered at the top of the stem; spikelets △a 1-flowered, purplish, unawned; flowering July—September. Sandy and waste places, often near the sea. B (rare), F, G. Map 191.

2. **Smooth Finger-grass** *Digitaria ischaemum*. Annual; stems to 40 cm, often bent at the base. Leaves flat with usually *hairless* sheaths; ligules *short, blunt*. Inflorescence a small cluster of 2–6 finger-like *racemes* (to 11 cm long) at the top of the stem; spikelets △b unawned, *2-flowered*, in groups of 3, the upper glume *equalling* the spikelet; flowering August—September. Casual in sandy fields, waste ground, often at ports. (B, F, G).

3. **Hairy Finger-grass** *Digitaria sanguinalis*. Differs from Smooth Finger-grass (2) in being sometimes taller (to 60 cm), with leaf-sheaths *hairy*, more (to 16) and longer (to 20 cm) racemes, the spikelets △c in *pairs* and the upper glume *shorter* than the spikelet; flowering August—October. Sandy fields, waste places. (T).

4. **Dogstooth Grass** *Dichanthium ischaemum*. Tufted and creeping perennial; stems to 100 cm. Leaves flat; ligules present. Inflorescence a fan of 3–15 finger-like *racemes*, to 7 cm long, clustered at the top of the stem; spikelets △d 2-flowered, *long-awned*, sometimes purplish; flowering July—August. Dry banks and fields. F, G.

5. **Maize** *Zea mays*. A well-known cereal, grown both for animal feed and as Sweetcorn; annual, with stems to 400 cm or more. Leaves *very broad*, to 12 cm. Inflorescences of 2 kinds: *male*, with pairs of finger-like racemes at the *top* of the stem; and *female*, with rows of spikelets on a stout axis, *enfolded* by a thick sheath and with protruding styles, at the base of the leaves *along* the stem; flowering July—September. A relic of cultivation, or on rubbish dumps and other waste ground. (B, F, G).

6. **Common Millet** *Panicum miliaceum*. A common constituent of bird-seed mixtures, sometimes cultivated for fodder; annual, with stems to 120 cm. Leaves flat, to 20 mm wide, with *hairy* sheaths. Inflorescence a stout stiff panicle, *nodding* at the tip; spikelets 2-flowered, *unawned*, the lower glume sharply pointed; flowering August—September. Waste ground, rubbish dumps. (B, F, G). Other *Panicum* species occur more rarely.

3-c

4-d

△ ×3

1-a

1a-b

1b-c

2-d

1. **Common Cord-grass** *Spartina anglica*. An often abundant, loosely tufted creeping perennial, commonly planted to stabilise coastal mud; stems stout, to 130 cm. Leaves *yellow*-green, flat or channelled, to *15 mm* wide; ligule a *ring of hairs*, 2–3 mm. Inflorescence a cluster, up to 35 cm long, of *3–6 spikes*, the axis ending in a 50 mm bristle *well above* the top spike; spikelets △a 1-flowered, unawned, the glumes hairy, anthers 8–12 mm; flowering July–November. Wet coastal mud. B, (F, G). Map 192. Common Cord-grass is a wholly new species, derived from chromosome doubling of **1a* Townsend's Cord-grass** *S.* × *townsendii*, the hybrid between Smooth and Small Cord-grasses (1b and 2). Townsend's has shorter ligule hairs (1–2 mm) and 2–4 flower-spikes (spikelet △b) with anthers 6–7 mm and the terminal bristle shorter (to 40 mm), but is best told by its lack of fertile pollen. Much less common than Common Cord-grass. B, F, (G). **1b* Smooth Cord-grass** *S. alternifolia*, naturalised from North America, has narrower (5–10 mm) green leaves, 5–13 flower-spikes (spikelet △c), terminal bristle to 25 mm and the glumes only sparsely hairy. Now almost extinguished by Common Cord-grass. (B, rare, F). Map 193.

2. **Small Cord-grass** *Spartina maritima*. Differs from Common Cord-grass (1) in its shorter stems (to 70 cm), narrower *green* or purplish leaves, with minute ligule hairs less than 1 mm, shorter (to 10 cm) cluster of only 2–3 flower-spikes (spikelet △d), anthers 4–6 mm and *short* terminal bristle, to 14 mm; flowering July–September. Saltmarshes on firmer mud. B, F, G. Map 194.

3. **Cut Grass** *Leersia oryzoides*. Creeping perennial; stems to 100 cm, *downy* at the joints. Leaves flat, pale yellowish-green, *rough-edged*, the sheaths roughish, *inflated*, often partly or even wholly *enfolding* the inflorescence; ligules very short. Inflorescence a panicle, spreading when freed from its sheath, the branches *wavy*; spikelets 1-flowered, unawned; flowering August–October. Freshwater margins, often scrambling over other plants. T, but B (rare). Map 195.

4. **Cockspur** *Echinochloa crus-galli*. Loosely tufted annual; stems to 100 cm. Leaves flat; *no ligules*. Inflorescence a panicle of one-sided racemes, to 25 cm, the longer ones with *short branches*; spikelets 2-flowered, sharply pointed or *shortly awned*; flowering August–October. Bare or waste ground, rubbish dumps. (B, F, G). Map 196. **4a*** *E. colonum* is smaller in most dimensions, with the inflorescences to 15 cm, the always unbranched racemes to 3 cm and the spikelets unawned. A rarer casual.

5. **Great Millet** *Sorghum halepense*. Widely cultivated as a cereal in warm climates; a stout *creeping* perennial with stems to 150 cm. Leaves long, flat, broad, with *ligules*. Inflorescence a panicle, the branches with racemes of 2-flowered spikelets, the lemmas with a *bent awn*; flowering August–September. Bare or waste ground, rubbish dumps. (B, F).

The Sedge Family Cyperaceae

See p. 11 for introduction to Sedge Family.

1. **Wood Club-rush** *Scirpus sylvaticus.* Creeping; stems *all flowering,* medium to tall, to 120 cm, *bluntly* 3-sided, smooth, *leafy.* Leaves 5–20 mm wide, flat, roughish on margins and midrib. Inflorescence a *loose* cluster of long-stalked clusters of 2–5 shorter-stalked, egg-shaped spikelets, with leaflike bracts about *as long as* the inflorescence; individual flowers with 6 *straight roughish* bristles (the residual petals) and *greenish* glumes; flowering May–July. Fruit yellow-brown. Wet woods and shady fens and marshes. T. Map 197. **1a** *S. radicans* □ is tufted, with non-flowering shoots creeping and rooting, solitary pointed spikelets, and longer, smooth, wavy bristles. Damp but less shady places. F, G, S. Map 198. **1b** *S. atrovirens* has spikelets in clusters of 8–20 and red-brown glumes. A North American plant becoming established in north-east France.

2. **Sea Club-rush** *Scirpus maritimus.* Creeping; stems *leafy,* medium to tall, to 120 cm, *sharply* 3-sided, roughish at top. Leaves 2–20 mm wide, keeled, often rough on margins and midrib. Inflorescence a *tight* cluster of *unstalked* egg-shaped spikelets, with leaf-like bracts much *longer* than the inflorescence; glumes red-brown, awned; flowering June–August. Fruit blackish-brown. Forms large stands in brackish water, mainly *by the sea.* T. Map 199.

3. **Round-headed Club-rush** *Scirpus holoschoenus.* Tufted; stems medium to very tall, to 150 cm, *rounded,* smooth, *leafless* or with a few short straplike leaves near the base. Inflorescence a loose cluster of stalked *globular* heads of small egg-shaped spikelets; glumes pale brown, shortly awned; flowering June–September. Fruit pale brown. Damp places; in west Europe in damp sandy places by the sea; in central Europe in wet meadows. B (rare), F, G. Map 200.

The Sedge Family Cyperaceae

△ ×3

1a

1: **Common Club-rush or Bulrush** *Scirpus lacustris*. Creeping; stems solitary, tall to very tall, to 300 cm, *rounded*, smooth, *green*, leafless above, but with narrow strap-like leaves under water. Inflorescence a loose cluster of stalked heads of unstalked egg-shaped spikelets, the lower bract *shorter* than the inflorescence; glumes red-brown, styles 3; flowering June—August. Fruit grey-brown. In slow-moving rivers and streams and around lakes, often in deep water. T. Map 201. **1a*** **Grey Club-rush** *S.l.* ssp. *tabernaemontani* is shorter, to 170 cm, and has glaucous stems, spikelet △a, glumes red-dotted and 2 styles. Also in brackish water. Map 202. **1b*** *S.* ×*carinatus*, the hybrid between Common and Triangular Club-rushes (1 and 3), may occur in the absence of the parents; its stems are 3-sided at the top.

2: **Sharp Club-rush** *Scirpus pungens*. Creeping; stems *solitary*, medium to tall, to 100 cm, *3-sided*, leafless except for a few strap-like leaves at their base. Inflorescence a compact head of *1–6* unstalked spikelets, the lower bract *longer* than the inflorescence; glumes red-brown, *blunt*, with 2 pointed *side lobes* and *no bristles*; flowering June—August. Fruit yellow-brown. Fresh and brackish marshes, sandy river banks. B (rare), F, G. Map 203. **2a** *S. mucronatus* □ is tufted, completely leafless and with more (2–20) spikelets, glumes minutely pointed, unlobed and with 6 bristles, and blackish-brown fruit. F, G. Map 204.

3: **Triangular Club-rush** *Scirpus triqueter*. Creeping; stems solitary, medium to very tall, to 150 cm, *3-sided*, leafless except for a single strap-like leaf at the base. Inflorescence a cluster of stalked and unstalked heads of egg-shaped spikelets, the lower bract as long as or *longer* than the inflorescence; glumes red-brown, unlobed, with 6 bristles; flowering June—September. Fruit red-brown. Muddy rivers and estuaries and other damp and wet places, often in brackish water. Hybridises with Common Club-rush (1) to give *S.* × *carinatus* (1b). B (rare), F, G. Map 205.

4. **Dwarf Club-rush** *Scirpus supinus*. *Annual*; stems *low to short*, to 30 cm, leafless. Inflorescence a single head of 2–12 egg-shaped spikelets close together, the bract much *longer* than the inflorescence; glumes red-brown; flowering July—September. Fruit blackish-brown. Damp places. F, G. Map 206.

The Sedge Family Cyperaceae

△ ×3

1-a

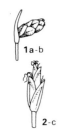

1a-b

2-c

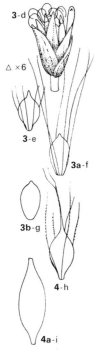

3-d

△ ×6

3-e

3a-f

3b-g

4-h

4a-i

1.* Bristle Club-rush *Scirpus setaceus.* Delicate tufted annual, or short-lived perennial; stems low to short, to 30 cm, prolonged *well beyond* inflorescence as a pin-like bract △a. Leaves thread-like, in pairs at base of stem. Inflorescence a single head of *1–4* egg-shaped spikelets; glumes purple-brown; flowering May–July. Fruit brown or black, shiny, conspicuous when ripe. Damp, sandy, usually rather bare places, often by fresh water. T. Map 207. **1a* Slender Club-rush** *S. cernuus* □ is smaller and slenderer, always annual, and with the bract only as long as or even shorter than the inflorescence △b, which usually consists of a single spikelet; glumes whitish or greenish; fruit red-brown, matt. Often in short turf near the sea. B, F. Map 208.

2.* Floating Club-rush *Scirpus fluitans.* Stems matted, to 50 cm or even longer, rooting at the joints, *floating* and largely submerged in still and slow-moving, clear, often fairly *acid water*; confusable with pondweeds *Potamogeton*, whose flowers have distinct sepal-like petals. Leaves 2 mm wide, submerged, fresh green. Inflorescence △c a *single* long-stalked egg-shaped spikelet, projecting above the surface; glumes greenish; flowering May–July. Fruit whitish or yellowish. T. Map 209.

3.* Deergrass *Scirpus cespitosus.* Tufted, sometimes forming conspicuous tussocks, but not creeping; stems low to short, to 35 cm, *rounded*, smooth, leafless except for a single *short strap-like leaf* (which distinguishes from the spike-rushes, p. 122) near the base. Inflorescence △d a *single*, more or less egg-shaped spikelet at the top of the stem; glumes yellowish to red-brown, with bristles a little *longer* than the greyish to yellow-brown matt fruit △e; flowering May–June. Often abundant on peaty moors, heaths and bogs on markedly *acid* soils. T. Map 210. **3a Alpine Deergrass** *S. hudsonianus* has stems 3-sided and rough at the top, and bristles much longer than the fruit △f; usually on base-rich peat. T, but extinct in B. Map 211. **3b Dwarf Deergrass** *S. pumilus* is shorter, with long runners and no or only a few very short bristles on the fruit. △g. Damp places on calcareous soils. S (rare). Map 212.

4.* Flat Sedge *Blysmus compressus.* Tufted or creeping; stems low to medium, to 45 cm, leafy at the base. Leaves 1–4 mm wide, *flat*, roughish. Inflorescence a *flattened* oblong head of 10–25 spikelets; glumes yellowish- or red-brown; flowering June–July. Fruit △h dark brown, shiny, with *long brown* bristles. *Fresh* marshes and damp grassland. T. Map 213. **4a* Saltmarsh Flat Sedge** *B. rufus* □ has the leaves smooth with inrolled margins, smaller heads with only 3–8 spikelets, and the bristles, when present, white and much shorter than the yellowish fruit △i. Saline marshes. B, G, S. Map 214.

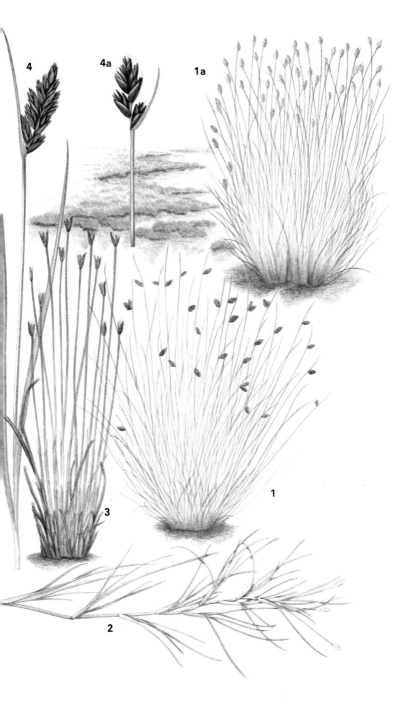

The Sedge Family Cyperaceae

COTTON-GRASSES or **BOG COTTON** *Eriophorum* look very different in spring, when the flowering heads have bright yellow anthers, compared with the fruiting heads in summer with their long white cottony hairs. Flowering April–May; fruiting May–July.

1: Common Cotton-grass *Eriophorum angustifolium.* Creeping; stems short to tall, to 75 cm, 3-sided *at the top.* Leaves dark green, often tinged reddish, especially in summer, 3–5 mm wide, grooved, with a *long* 3-sided tip and a *short ligule.* Inflorescence a loose drooping cluster of 3–7 *smooth*-stalked spikelets. Fruit dark brown, the white cottony hairs unbranched. Bogs and other wet, peaty and mainly acid places, often whitening the ground and a useful sign of where it is unsafe to walk; frequently in standing water. T. Map 215.

△ ×2

2: Broad-leaved Cotton-grass or **Fen Cotton-grass** *Eriophorum latifolium.* Loosely *tufted*; stems short to tall, to 70 cm, 3-sided *throughout.* Leaves 3–8 mm wide, with a *short* 3-sided tip and *no ligule.* Inflorescence a loose drooping cluster of 2–12 spikelets △a with *rough 3-sided* stalks. Fruit red-brown, the cottony hairs *branched* at the tip. Fens and marshes with *base-rich* water. T. Map 216.

2-a

3: Slender Cotton-grass *Eriophorum gracile.* Creeping; stems medium, to 60 cm, slender, 3-sided *throughout.* Leaves narrow, 1–2 mm wide, with a *short ligule.* Inflorescence a loose drooping cluster of 3–6 spikelets △b with *rough 3-sided* stalks. Fruit yellow-brown, the cottony hairs not branched. Bogs and other wet places on *acid* soils, often growing with Slender Sedge *Carex lasiocarpa* (p. 140). T, but B (rare). Map 217.

3-b

4: Harestail Cotton-grass *Eriophorum vaginatum.* Tufted, forming *tussocks*; stems medium, to 60 cm, 3-sided *at the top.* Leaves thread-like, *c.* 1 mm wide, 3-sided. Inflorescence a *single* erect spikelet, emerging from a conspicuously inflated sheath; glumes grey, anthers *c.* 3 mm. Fruit yellow-brown. Wet moors, heaths and bogs, on *acid* peaty soils. T. Map 218. **4a** *E. brachyantherum* has smaller tussocks, darker glumes, cottony hairs often tinged brown, and shorter anthers. S. Map 219. **4b** *E. russeolum* □ is creeping, with rounded stems, darker glumes, red-brown cotton and fruit beaked. S. Map 220. **4c** *E. scheuchzeri* is creeping and has rounded stems, slightly broader leaves, 3-sided only at the tip, darker glumes and anthers *c.* 1 mm. S. Map 221. The last three species grow largely on tundra and in alpine bogs.

6: Many-stalked Spike-rush *Eleocharis multicaulis* (see p. 122). *Tufted*; stems slender, short, to 30 cm, the top sheath △f *pointed* and *very obliquely* truncated, the lower sheaths yellowish- or pale brown. Spikelet 5–15 mm, pointed, brown, with 10–30 often *viviparous* flowers, the lowest glume usually *c.* one-quarter as long as the spikelet; flowering June–August. Fruit yellow- or olive-brown with longer bristles. Pools in bogs and on wet heaths on *acid* soils. T. Map 229.

The Sedge Family Cyperaceae

SPIKE-RUSHES *Eleocharis* have single spikelets at the top of hairless un-branched erect stems. Except where stated, they are perennials and have brownish sheaths instead of leaves, the lowest glume blunt, *flowerless* and *encircling* the base of the spikelet, and 3 styles. Deergrass (p. 188) is similar, but its top sheath has a tiny leaf.

△ ×6

1. **Few-flowered Spike-rush** *Eleocharis quinqueflora.* Tufted, with underground runners; stems low to short, to 30 cm, the top sheath △a *obliquely* truncated. Spikelet 4–10 mm, brown, with 3–7 flowers; lowest glume sometimes *with a flower* and *more than half* as long as spikelet; flowering June–July. Fruit black, drying grey. Bare places in marshes and fens, usually peaty and base-rich. T. Map 222.

2. **Dwarf Spike-rush** *Eleocharis parvula.* Tufted, with whitish thread-like underground runners ending in *whitish tubers*; stems thread-like, *low*, to 8 cm. Leaves *thread-like*, lax. Spikelet 2–3 mm, *greenish*, with 3–5 flowers; flowering August–September. Fruit yellow. Tidal mud or by salt lakes inland, often forming non-flowering patches. T. Map 223.

3. **Needle Spike-rush** *Eleocharis acicularis.* Creeping; stems *4-angled*, up to 50 cm when submerged and not flowering, often only to 10 cm when flowering on mud; top sheath △b *squarely* truncated and slightly inflated. Spikelet 2–4 mm, dark brown, with 3–11 flowers, the lowest glume sometimes *with a flower* and *less than half* as long as spikelet; flowering June–October. Fruit pale brown or whitish. In seasonally dry shallow fresh water and on nearby mud, often making bright green 'lawns' and only flowering when exposed. T. Map 224.

4. **Common Spike-rush** *Eleocharis palustris.* Creeping, with *small* tufts; stems *not* cracking easily or ridged when dry, short to medium, to 60 cm, the top sheath △c *squarely* truncated, yellow-brown, the lower sheaths *reddish*. Spikelet 3–20 mm, yellowish- to dark brown, with 20–70 flowers, the 2 lowest glumes short, with no flower and each *half*-encircling base of spikelet; *styles 2*; flowering May–August. Fruit yellow to brown. Marshes, damp grassland and shallow fresh water. T. Map 225. **4a** Northern Spike-rush *E. austriaca* (incl. *E. mamillata*) has easily cracked stems, ridged when dry, lower sheaths not reddish and compacter spikelets; usually on base-rich soils. Maps 226 and 227. **4b** Oval Spike-rush *E. ovata* is a tufted annual, with the top sheath △d greenish, very obliquely trun-cated and sometimes with a tiny leaf-blade, lower sheaths purp-lish, and spikelet shorter, more egg-shaped and red-brown; bare ground, wet in winter. F, G.

5. **Slender Spike-rush** *Eleocharis uniglumis.* Creeping; stems slender, short to medium, to 60 cm, not ridged when dry; top sheath △e *squarely* truncated, lower sheaths red-purple. Spikelet 5–12 mm, dark brown, with 10–30 flowers more widely spaced than other spike-rushes, the lowest glume en-circling the whole spikelet; *styles 2*; flowering May–August. Saline or base-rich marshes. T. Map 228.

6. **Many-stalked Spike-rush** *Eleocharis multicaulis.* See p. 120.

1-a

3-b

4-c

4b-d

5-e

6-f

The Sedge Family Cyperaceae

1. **Galingale** *Cyperus longus* (incl. *C. badius*). Creeping; stems medium to *very tall*, to 150 cm, 3-sided. Leaves 2–10 mm wide, glossy, rough-edged. Inflorescence a compound *umbel* (distinguishing from Wood Club-rush (p. 114), whose flowers are in a forking cluster, with olive-brown glumes and yellowish fruit) with up to 10 rays, each with 4–25 spikelets, the outer bract longer than the inflorescence; glumes dark or *red-brown*; stamens 3; flowering August–September. Fruit *brownish-black*. Damp places and by fresh water. B, F. Map 230.

2. **American Galingale** *Cyperus eragrostis* (*C. vegetus*). Creeping; stems medium to tall, to 90 cm, 3-sided. Leaves 2–10 mm wide. Inflorescence a simple or compound umbel with 8–10 rays, each with a *close head* of 8–13 spikelets, the bracts *all* longer than the inflorescence; glumes *yellowish*, stamen *1*; flowering August–October. Fruit *grey*. A tropical American plant naturalised here and there in the south of the region. (B, F). Other *Cyperus* spp. are occasional casuals, among them *C. distachyos*, *C. dives*, *C. rotundus*, *C. serotinus* and *C. virens*.

3. **Brown Galingale** *Cyperus fuscus*. A tufted *annual*; stems low to short, to 30 cm, 3-sided. Leaves 2–7 mm wide. Inflorescence a simple or compound umbel of 3–8 rays, each with a *close head* of spikelets, the bracts longer than the inflorescence; glumes dark or *red-brown*; flowering July–September. Fruit *white*. *Bare mud*, especially in dried-up ponds. T, but B (rare). Map 231. **3a Yellow Galingale** *C. flavescens* may be taller and have narrower leaves and either a loose head of unstalked spikelets or a simple umbel with 1–4 rays; glumes yellowish; fruit blackish-brown. Also on bare peat; decreasing. F, G. Map 232.

4. **Great Fen Sedge** *Cladium mariscus*. Creeping and *patch-forming*; stems stout, hollow, rounded, tall to *very tall*, to 250 cm. Leaves long, up to 200 cm, but characteristically bent in an inverted V and so appearing much shorter; also broad, to 10–15 mm wide, glaucous, stiff and *saw-edged*. Inflorescence a *panicle*, each cluster with 5–10 egg-shaped spikelets; glumes pale brown; flowering July–August. Fruit dark brown, shiny. Fens, swamps, lake shores, persisting into alder carr, usually in base-rich water, sometimes in dense stands. Still cut for litter in East Anglia, where known as 'sedge', a name also perpetuated in Sedgemoor, Somerset. T. Map 233.

The Sedge Family Cyperaceae

1. **White Beak-sedge** *Rhynchospora alba.* Loosely tufted; stems short to medium, to 40 cm, rounded or sometimes angled at the top. Leaves 1–2 mm wide, pale green, grooved, the lower often with *bulbils* at the base, nearly *as long as* the stems. Inflorescence a close *flattish* cluster of 2-flowered spikelets, *whitish* at first, becoming pale brown, often with 1–2 smaller stalked heads lower down; the top bract *as long as* or a little longer than the inflorescence; flowering June–September. Decreasing in bogs and wet peaty heaths and moors on acid soils. T. Map 234.

2. **Brown Beak-sedge** *Rhynchospora fusca.* Creeping, stems short, to 30 cm, rounded or sometimes angled at the top. Leaves *c.* 1 mm wide, with *no bulbils*, much *shorter* than the stems. Inflorescence an *egg-shaped* head of red-brown spikelets, often with another stalked one below, the top bract *much longer* than the inflorescence; flowering May–July. Wet peaty heaths and bog pools on acid soils. T. Map 235.

3. **Black Bog-rush** *Schoenus nigricans.* Tufted; stems medium, to 60 cm. Leaves 1–2 mm wide, grey-green, with inrolled margins, their basal sheaths *blackish.* Inflorescence a head of 5–10 *blackish*-brown spikelets, each with 1–4 flowers, the lower bract *longer* than the inflorescence; flowering May–July. Fruit whitish. *Fens* on base-rich soils, dune slacks, peaty flushes and, in the far west, bogs. T. Map 236.

4. **Brown Bog-rush** *Schoenus ferrugineus.* Shorter and slenderer than Black Bog-rush (3), and with shorter leaves, their basal sheaths *red-brown,* the inflorescence with only 1–3 spikelets and so *narrower,* and the lower bract only *as long as* the inflorescence; flowering April–July. Flushes and other wet peaty spots on base-rich or slightly acid soils; apparently often overlooked. T, but B (rare). Map 237.

5. **False Sedge** *Kobresia simpliciuscula.* Tufted; stems low to short, to 20 cm, stiff, 3-sided. Leaves *c.* 1 mm wide, grooved, short, with basal sheaths *leafy,* pale orange-brown and matt. Inflorescence an elongated head with 3–10 short spikes of 4–8 small dark red-brown *1-flowered* spikelets, each spike male above and female at the base. Fruit pale brown, shiny, distinguished from all true sedges *Carex* (pp. 128–166) by the nut being only *partly* enveloped by 1 of the 2 glumes at its base; flowering June–July. *Damp,* often rather bare places on calcareous soils. B, S. Map 238. **5a** *K. myosuroides* □ has the basal sheaths leafless, brown and shiny, longer leaves and a single spike with 8–10 2-flowered spikelets, each with a male flower above a female, except for the top spikelet, which has several male flowers; drier places. S. Map 239.

The Sedge Family Cyperaceae

True **SEDGES** *Carex* (pp. 128–166) are superficially grass-like *perennials*, especially in their leaves, which are usually wintergreen, but their stems are always *solid* and often 3-sided and have *no joints* at the leaf-junctions. The individual flowers are arranged all round the stem (not in rows like the grasses) in one or more *tight* panicles, here called *spikes*. They are petalless and sepalless, but each have a scale (*glume*) at the base; male and female flowers are *separate*, but (except for the 4 species on p. 136) in the same inflorescence. Female glumes usually have both a pale or green midrib and a pale margin. Most species flower, i.e. their conspicuous yellow anthers appear, in May or June. Fruits are small nuts, either 2- or 3-sided; ripe fruit is valuable for the identification of most species and essential for some.

1ː Greater Tussock Sedge *Carex paniculata*.   Easily identified throughout the year by its *substantial tussocks*, whose base of fibrous rhizomes may be up to 100 cm high and 100 cm wide. Stems to 150 cm, rough, with 3 flat sides and *dark brown* scales at the base. Leaves to 120 cm, rough, *dark green*, 5–7 mm wide. Inflorescence 5–15 cm, oblong, brown. Fruit △a ribbed, with a winged beak. On peaty soils, in fens, marshes and wet woods; typically with both Pond Sedges (p. 142). T. Map 240.

2ː Fibrous Tussock Sedge *Carex appropinquata*.   Best distinguished from Greater Tussock Sedge (1) by its much smaller tussocks, *shorter* stems to 80 cm, with *blackish* fibrous scales at the base, much *narrower* (1–2 mm) *yellowish-green* leaves, shorter (4–6 cm) reddish-brown inflorescence, and unwinged fruit △b. Similar places to Greater Tussock Sedge, sometimes on rather more acid peat. T. Map 241.

3ː Lesser Tussock Sedge *Carex diandra*.   Tufted, often with a creeping rhizome producing tufts at intervals; stems to 60 cm, rough, with 3 *convex* sides, the basal sheaths *greyish* or blackish-brown. Leaves 1–2 mm wide, rough, *greyish-green*. Inflorescence 1–5 cm, dark brown. Fruit △c unwinged, ribbed. Wet peaty soils, often fairly *acid*, including old meadows and alder carr. T. Map 242.

△ ×6

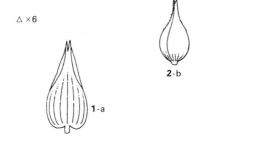

2-b

1-a

3-c

1

2

3

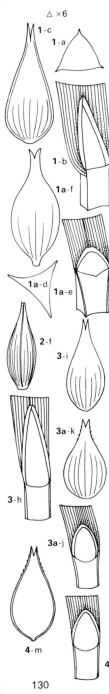

△ ×6

1-c
1-a
1-b
1a-f
1a-d
1a-e
2-f
3-i
3a-k
3-h
3a-j
4-m
4.

The Sedge Family Cyperaceae

1: **False Fox Sedge** Carex otrubae. Tufted, with stems to 100 cm, rough at the top, with 3 flat *unwinged* sides △a. Leaves up to 60 cm, 4–10 mm wide, with *auricles* at the base and *pointed* ligules △b to 10 mm. Inflorescence 3–7 cm, *greenish* or brown, often with a long lower bract. Fruit △c. *Common* in damp grassy places, often on heavy soils. Hybridises with Remote Sedge (p. 136). T. Map 243. **1a* True Fox Sedge** C. vulpina is somewhat stouter, with sharply angled, winged stems △d with concave sides; leaves with shorter blunt ligules △e, overlapping margins and no auricles; and red-brown inflorescences with shorter bracts, flowering a week or two earlier. Fruit △f. Wetter places. Map 244. **1b* American Fox Sedge** C. vulpinoidea, smaller, with awned bracts and flowers, occurs as a casual.

2: **Elongated Sedge or Gingerbread Sedge** Carex elongata. Forms floppy tussocks; stems rough, to 80 cm. Leaves flat, 2–5 mm wide, pale *yellow-green*, turning red-brown in autumn; ligule long, *pointed*. Inflorescence 3–10 cm, *branched*, dark reddish-brown, with 5–15 spikes, each *female above* and male below, or the lower all female. Fruit △g spreading when ripe. Local in damp meadows and alder woods, and by fresh water. T. Map 245.

3: **Spiked Sedge** Carex spicata. Tufted, with stems up to 100 cm, but usually much shorter; basal sheaths often *purple-tinged*. Leaves 2–4 mm wide, flat, leaf-bases and sometimes bracts often tinged *wine-red*; ligules △h long, *pointed*. Inflorescence 2–3 cm, greenish or brownish, with 3–10 spikes and awned glumes. Fruit △i with a *long narrow beak*. Grassy, sometimes dampish, mainly calcareous places. T. Map 246. **3a* Prickly Sedge** C. muricata ssp. lamprocarpa has no purple or vinous tinge, shorter blunter ligules △j, pale glumes and more globular fruit △k, narrowing gradually to the beak. Widespread on acid soils. Map 247. **3b*** C.m. ssp. muricata has dark glumes and fruit abruptly narrowed to the short beak, otherwise like ssp. lamprocarpa; rare in limestone grassland. Map 247.

4: **Grey Sedge** Carex divulsa. Tufted; stems to 75 cm. Leaves 2–3 mm wide, flat or grooved; ligules △l broad, blunt. Inflorescence *long*, 5–15 cm, with 5–8, more *widely spaced* spikes than Spiked Sedge (3). Fruit △m gradually narrowed to the beak. Grassy places, often on calcareous soils. A variable species, separated into 2 subspecies, poorly defined in south England: *divulsa* with dark green leaves, mature fruit blackish and flowering June–October; and *leersii*, stiffer, with yellow-green leaves, a compacter inflorescence, larger red-brown fruit and flowering May–August. T. Map 248.

Fruits and stems △ ×6
Ligules △ ×4

1

4

2

3

The Sedge Family Cyperaceae

△ ×6

1.* **Sand Sedge** *Carex arenaria*. Far-creeping and easily told by the *long lines* of spaced shoots arising from its long rhizomes, especially conspicuous on bare sand; stems to 40 cm, often *curved*. Leaves 2–4 mm wide, often inrolled, *equalling* stems; ligules pointed. Inflorescence oblong, 3–8 cm, with 5–18 spikes, the *upper male*, the middle mixed and the lower female; female glumes pale red-brown; flowering May. Fruit △a yellow-brown, beaked. Coastal *dunes* and other sandy places. T. Map 249. **1a French Sedge** *C. ligerica* □ is smaller and slenderer, with much shorter rhizomes; narrower (1–2 mm) leaves shorter than stems; inflorescence shorter (2–3 cm) with female spikes above and male below, female glumes dark-brown and flowering April. Fruit △b. F, G, S (rare). Map 250. **1b** *C. reichenbachii* may be taller and has leaves 2–3 mm, longer than stems, inflorescence 3–5 cm, female spikes above and male below, female glumes pale yellow-brown and fruit △c greenish. Intermediate in many respects between Sand and Quaking-grass Sedges (1 and 4). Pinewoods and sandy heaths. F, G. Map 251.

1 -a

1a-b

2.* **Brown Sedge** *Carex disticha*. Far-creeping, with erect stems to 100 cm. Leaves 2–4 mm wide, *shorter* than stems; ligules blunt. Inflorescence oblong, 3–7 cm, with 15–30 spikes, the upper and lower female, the *middle male*; female glumes red-brown. Fruit △d red-brown, beaked. Fens, *damp grassland* and dune slacks, where it may overlap with Sand Sedge (1). T. Map 252.

3. **Early Sedge** *Carex praecox*. Far-creeping, with stems to 30 cm. Leaves *very narrow*, 1–2 mm, *much shorter* than stems. Inflorescence oblong, *short*, c. 2 cm, with 3–7 spikes, the upper female, the lower male; female glumes red-brown; flowering *April*-May. Fruit △e red-brown, shortly beaked. Dry, often sandy, bare or grassy places. F, G. Map 253.

1b-c

4. **Quaking-grass Sedge** *Carex brizoides*. Far-creeping, with stems to 50 cm. Leaves 2–3 mm wide, *longer* than stems. Inflorescence narrow oblong, slightly curved, 2–3 cm, with 5–8 spikes, the upper female, the lower male; female glumes *pale brown to greenish*. Fruit △f grey-green, long-beaked. Damp broad-leaved woods and other shady places. F, G. Map 254.

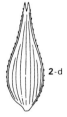

2-d

3-e

4-f

The Sedge Family Cyperaceae

1. **String Sedge** *Carex chordorrhiza*.   Far-creeping, with stems to 40 cm, smooth. Leaves narrow, 1–2 mm wide; ligules short, rounded. Inflorescence *very short*, 7–15 mm, with 2–4 spikes, each male above and female below, or the lower all female; female glumes red-brown. Fruit △a yellowish- to dark brown. *Wet bogs*. T, but rare in B, F. Map 255.

2. **Divided Sedge** *Carex divisa*.   Creeping, with stems to 80 cm, rough at the top. Leaves narrow, 2–3 mm wide, wintergreen; ligules short, blunt. Inflorescence distinctly *lobed*, 1–3 cm, with 3–8 spikes, usually overtopped by a narrow bract, each male above and female below, or the lower all female; female glumes purple-brown. Fruit △b. Damp grassy or marshy places, dune slacks, often *near the sea*. B, F, G. Map 256.

3. **Curved Sedge** *Carex maritima*.   Far-creeping, often half-buried in sand; stems to 18 cm, often markedly *curved*. Leaves thread-like, to 1 mm wide, often wintergreen. Inflorescence short, 1–2 cm, almost globular, with 4–8 spikes, the *upper all male*, the lower female, or sometimes male at the top; female glumes dark brown. Fruit △c blackish-brown when ripe, disappearing after mid-July. *Coastal*, mainly on sand, but with a freshwater supply; also on some mountains on the Continent. B, G, S. Map 257.

4. **Oval Sedge** *Carex ovalis*.   *Tufted*; stems to 60 cm, rough at the top, often curved. Leaves 1–3 mm wide, rough-edged, wintergreen; ligules blunt. Inflorescence *compact*, 2–4 cm, with 2–9 spikes, each *female above* and male below, or the lower all female; female glumes rufous. Fruit △d pale brown. Damp grassland, woodland rides, heaths and moors on *acid soils*. T. Map 258.
4a *C. macloviana* □ is shorter, with inflorescence 1–2 cm, fewer, more crowded spikes; female glumes and fruit dark brown. Grassy places in north Scandinavia. Map 259.

△ ×6

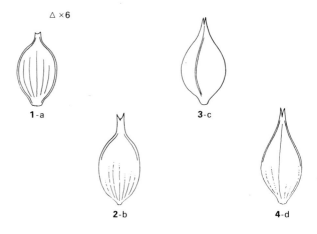

1-a

3-c

2-b

4-d

The Sedge Family Cyperaceae

1. **Remote Sedge** *Carex remota.* Tufted; stems to 60 cm. Leaves pale green, 1–2 mm wide. Inflorescence 8–20 cm, the lower flower spikes *well spaced out* down the stem and having *long leaf-like bracts*; flower spikes 3–10 mm, each male at top, female below. Fruit △a green. Hybridises with False Fox Sedge (p. 130). *Damp* woods and other shady places. T. Map 260.

2. **Bohemian Sedge** *Carex bohemica.* Tufted; stems to 30 cm. Leaves *c.* 2 mm wide. Inflorescence 2 cm *across*, almost *globular*, with leaf-like bracts, the upper *spreading*, the lower erect; flower spikes *grass-green* with whitish female glumes. Fruit △b pale green to pale yellow, long-beaked. Damp places, often by fresh water, dried lake-beds; rarely persisting. F, G. Map 261.

3. **Star Sedge** *Carex echinata.* Tufted; stems to 40 cm. Leaves 1–2 mm wide. Inflorescence 1–3 cm, the flower spikes 3–6 mm, fairly close together, the top one female above and male below, the rest all female; female glumes red-brown. Fruit △c greenish- to dark brown, spreading *star-wise*. Wet heaths, bogs, marshes, on *acid soils*. T. Map 262.

4. **Dioecious Sedge** *Carex dioica.* One of only 4 sedges in the region with *male and female flowers on separate plants.* Creeping; stems to 40 cm, rounded. Leaves thread-like, 0.3–1 mm wide. Inflorescence 5–20 mm, with a single male or female flower spike, the female shorter and thicker than the male; glumes red-brown. Fruit △d red- to purple-brown, *abruptly* narrowing to a short beak. Calcareous and other base-rich marshes, preferring silty to peaty soils. T. Map 263. **4a** *C. parallela* has the fruit gradually narrowing to the beak. S. Map 264. **4b** *C. davalliana* forms conspicuous tussocks and has the fruit △e gradually narrowing to the beak. B (extinct), F, G. Map 265. **4c** *C. scirpoides* is loosely tufted, with shorter, bluntly 3-sided stems, broader leaves, female glumes blackish and fruit yellow-brown; wet grassland in mountains in Arctic Norway only.

△ ×6

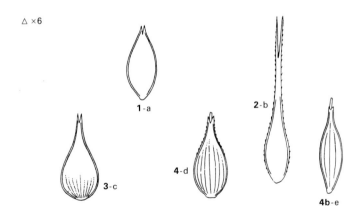

1-a

2-b

4-d

3-c

4b-e

1

2

3

4

♀   ♂

The Sedge Family Cyperaceae

△ ×6

1. **Haresfoot Sedge** *Carex lachenalii.* Tufted, often loosely; stems to 30 cm, *bluntly* 3-sided. Leaves 1–2 mm wide, *green.* Inflorescence short, 1–2 cm, with 3–4 spikes, *each female above and male below;* female glumes dark red-brown; flowering July–August. Fruit △a beaked, greenish-, yellowish- or dark brown. Marshy grassland, wet heaths, *mountain* ledges and flushes. B, S. Map 266. **1a Peat Sedge** *C. heleonastes* has sharply 3-sided stems to 40 cm, bluish-green leaves, paler female glumes and shorter, pale to dark brown fruit △b; wet peaty places. G, S. Map 267.

2. **Saltmarsh Sedge** *Carex glareosa.* Tufted; stems to 20 cm, bluntly 3-sided. Leaves thread-like, 0.5 mm, grooved, *greyish.* Inflorescence short, 1–2 cm, with 2–3 spikes, the top one female above and male below, the *lower all female;* female glumes dark red-brown. Fruit △c greenish- to reddish-brown, beaked. Saltmarshes and rocks by the sea. S. Map 268. **2a** *C. mackenziei* is much taller (to 40 cm), often loosely tufted and creeping and has broader (2–3 mm) flat, often yellow-green leaves, longer (2–5 cm) inflorescences of 3–6 spikes, paler or yellowish female glumes and greyer, short-beaked fruit △d. Hybridises with White Sedge (3). Map 269.

3. **White Sedge** *Carex curta.* Tufted; stems to 50 cm, *sharply* 3-sided. Leaves 2–3 mm wide, *pale* green. Inflorescence 3–5 cm, with 4–7 spikes, *each female above and male below;* female glumes *whitish.* Fruit △e pale greenish or yellowish, *gradually* narrowed to a beak. Wet heaths, acid mires, often on mountains. T. Map 270. **3a** *C. lapponica* has narrower (1–1.5 mm) leaves, shorter inflorescences and the fruit △f abruptly narrowing to the beak; especially in bogs. S. Map 271. **3b** *C. brunnescens* is like *C. lapponica* but with greener leaves and darker brown glumes; in damp rather than wet places. G, S. Map 272.

4. **Darnel Sedge** *Carex loliacea.* Tufted and creeping; stems to 40 cm, slender. Leaves 1–2 mm wide, *bright* green. Inflorescence 1–3 cm, with 3–5 spikes, the lower *c. 1 cm apart, each female above and male below;* female glumes whitish-brown. Fruit △g greenish or brownish, not beaked, spreading when ripe. Damp woods and heaths, bogs. S. Map 273. **4a** *C. tenuiflora* has greyer leaves and very short (0.5–1 cm) compact inflorescences. Map 274.

5. **Fine-leaved Sedge** *Carex disperma.* Loosely tufted and creeping; stems to 50 cm, lax. Leaves *thread-like,* 1 mm wide, weak, *bright* green. Inflorescence 2–4 cm, very lax, with 2–6 spikes, each *male above* and female below, or the top one all male; female glumes whitish-brown. Fruit △h yellowish-green to dark brown, abruptly contracted to a short beak, spreading when ripe. Damp woods and heaths. S. Map 275.

1-a

1a-b

2-c

2a-d

3-e

3a-f

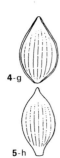

4-g

5-h

The Sedge Family Cyperaceae

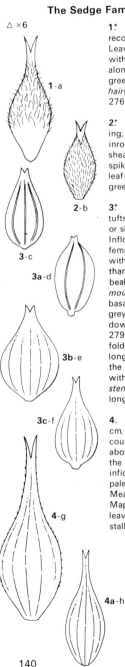

△ ×6

1-a

2-b

3-c

3a-d

3b-e

3c-f

4-g

4a-h

1. **Hairy Sedge** *Carex hirta.* One of the easiest sedges to recognise, by its *hairy leaves.* Tufted, creeping; stems to 70 cm. Leaves 2–5 mm wide, hairy, at least on the sheaths. Inflorescence with 2–3 male spikes above 2–3 female spikes, *all spaced out* along more than half the stem, the female spikes with pale greenish-brown glumes and long leaf-like bracts. Fruit △a *hairy,* green. Common in grassy places, usually damp. T. Map 276.

2. **Slender Sedge** *Carex lasiocarpa.* Loosely tufted, creeping; stems to 120 cm. Leaves *narrow,* 1–2 mm wide, the margins inrolled, *grey-green,* the lower with purplish- or *reddish-brown* sheaths. Inflorescence with 1–3 male spikes above 1–3 female spikes, the female spikes with purple-brown glumes and long leaf-like bracts; often very shy-flowering. Fruit △b *downy,* grey-green. Wet mires and reedswamps. T. Map 277.

3. **Russet Sedge** *Carex saxatilis.* Creeping, with small tufts; stems to 40 cm, sometimes curved. Leaves green, *equalling* or slightly shorter than stems, 2–4 mm wide, folded or grooved. Inflorescence with 1 male spike above 1–2 female spikes, the female spikes scarcely stalked (sometimes the lower stalked), with purple-brown glumes and the lower leaf-like bract *shorter* than the inflorescence. Fruit △c *abruptly* contracted to the short beak (0.5 mm), spreading when ripe. Mires and flushes on *mountains.* B, S. Map 278. **3a** *C. rotundata* has more rigid stems; basal sheaths often pale brown; shorter, narrower (1–2 mm), greyer, more rigid leaves; the lowest bract usually spreading or down-turned; and fruit △d inflated; on more acid soils. S. Map 279. **3b** *C. stenolepis* may be taller and has longer leaves, flat or folded; larger, always stalked female spikes; the lowest bract longer than the inflorescence; and fruit △e gradually narrowed to the beak, in bogs. S. **3c.** *C.* × *grahamii,* the hybrid of *C' saxatilis* with either *C. vesicaria* (p. 144) or *C. rostrata* (p. 144), is like *C. stenolepis* but with the fruit △f more abruptly contracted to a longer beak. B, S.

4. **Barley Sedge** *Carex hordeistichos.* Tufted; stems to 30 cm, the basal sheaths red-brown. Leaves 3–5 mm wide, glaucous, *longer* than stems. Inflorescence with 2–3 male spikes above 2–4 female spikes, the upper female spikes *clumped* but the lowest separate and with a leaf-like bract *longer* than the inflorescence; female glumes pale greenish-brown. Fruit △g pale yellowish- or orange-brown, shiny, arranged *in rows.* Meadows, ditches, waysides and damp grassy places. F, G. Map 280. **4a** **Rye Sedge** *C. secalina* is slenderer, with narrower leaves, all female spikes spread out down the stem and longer-stalked, and fruit △h not shiny or in rows. G. Map 281.

The Sedge Family Cyperaceae

△ ×6

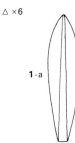

1-a

1-b

1a-c

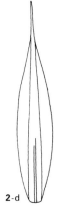

1: **Lesser Pond Sedge** *Carex acutiformis.* Creeping, usually in extensive patches; stems to 150 cm, sharply 3-sided, roughish. Leaves *broad*, 7–10 mm wide, *glaucous* when young, reddening in autumn, usually longer than stems; ligules blunt. Inflorescence with 2–3 male spikes above 3–4 female spikes, the upper female spikes *unstalked* and often male at top, the lowest with a short stalk; male glumes △a purple-brown, *bluntly* pointed; leaf-like bracts longer than the inflorescence. Fruit △b green, *abruptly* narrowed to the beak, usually with *3 styles* (cf. Slender Tufted Sedge (p. 164) which has 2). Marshy meadows, swamps and by rivers and ponds; common. T. Map 282. **1a Nodding Pond Sedge** *C. melanostachya* has shorter (to 50 cm) slenderer stems, narrower (2–4 mm) leaves with inrolled margins, somewhat shorter bracts, male flowers with sharply pointed glumes and fruit △c gradually narrowed to the beak. F, G. Map 283.

2: **Greater Pond Sedge** *Carex riparia.* Creeping, often in extensive patches; stems to 130 cm, sharply 3-sided, rough. Leaves *very broad*, 6–15 mm wide, *glaucous* when young, usually longer than the stems; ligules blunt. Inflorescence with 3–6 male spikes above 1–5 female spikes, the upper female spikes scarcely stalked and often male at top, the lower stalked, often *hanging down*; male glumes △d dark brown, *sharply* pointed, often bristle-tipped; leaf-like bracts longer than the inflorescence. Fruit △e green or brown, *gradually* narrowed to the beak. Marshy meadows, swamps and by ponds and rivers; common. T. Map 284.

3: **Cyperus Sedge or Hop Sedge** *Carex pseudocyperus.* Loosely tufted; stems to 90 cm, sharply 3-sided, rough at edges. Leaves to 120 cm, taller than stems, 5–12 mm wide, bright *yellow-green.* Inflorescence with 1 male spike above a cluster of 3–5 sausage-like female spikes, long-stalked and *hanging down*, the leaf-like bracts much longer than the inflorescence; female glumes △f with a *long awn.* Fruit △g green, spreading, soon falling. In and by swamps and still and slow-moving fresh water. T. Map 285.

2-e

3-f

3-g

2-d

3

1

2

The Sedge Family Cyperaceae

1. Bottle Sedge *Carex rostrata*.   Creeping, usually in extensive patches; stems to 100 cm, *bluntly* 3-sided and rough *above*. Leaves longer than stems, 2–7 mm wide, *greyish* above, sometimes with inrolled margins, wintergreen; ligules blunt. Inflorescence with 2–4 male spikes above 2–5 short-stalked female spikes, the lower bracts equalling or overtopping the inflorescence; glumes purple-brown, the female bluntly pointed. Fruit △a *egg-shaped*, yellow-green, matt. On wet base-poor peat, in swamps and around lake shores and margins. T. Map 286.

2. Bladder Sedge *Carex vesicaria*.   Creeping and patch-forming; stems to 120 cm, *sharply* 3-sided *throughout*. Leaves longer than stems, 4–8 mm wide, *green* or yellow-green, rough-edged, not overwintering; ligules pointed. Inflorescence with 2–4 male spikes above 2–3 female spikes, the lower spikes often well stalked, usually drooping and with a bract longer than the inflorescence. Fruit △b *narrower* than Bottle Sedge (1), olive-green, shiny. On wet peat and by fresh water, frequent around upland lakes. T. Map 287. **2a** *C. rhynchophysa* has leaves 8–15 mm wide and often shorter than stems, the lower female spikes shorter-stalked, and the fruit △c shorter, more globular and greener. S. Map 288.

3. Pendulous Sedge *Carex pendula*.   A distinctive and decorative sedge, growing in *substantial clumps*; stems to 180 cm (even to 240 cm), bluntly 3-sided. Leaves shorter than stems, *broad*, 15–20 mm wide, rough-edged, green or yellow-green above, glaucous beneath; ligules pointed. Inflorescence *drooping* gracefully, with 1–2 male spikes above 4–5 *long* (7–16 cm), narrowly cylindrical female spikes, the lower bracts shorter than the inflorescence; female glumes red-brown. Fruit △d grey-green, becoming brown. Frequent in woods on *heavy soils* and by shady streams. T. Map 289.

△ ×6

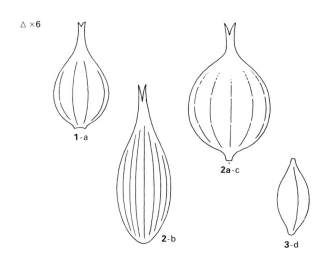

1-a

2a-c

2-b

3-d

The Sedge Family Cyperaceae

△ ×6

1-a

2-b

3-c

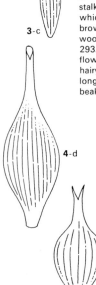

4-d

4a-e

1. **Wood Sedge** *Carex sylvatica.* Tufted; stems to 60 cm. Leaves rather *broad*, 3–6 mm wide, *flaccid*, shiny, green to yellow-green, wintergreen; ligules short, *blunt*. Inflorescence nodding, with 1 male spike above 3–5 *long-stalked* female spikes, the lower bracts sometimes overtopping the inflorescence; female glumes pale yellow or brown. Fruit △a green, *long-beaked*. Common in woods. T. Map 290.

2. **Hair Sedge** *Carex capillaris.* Tufted; stems to 40 cm, slender; leaves *narrow*, 1–2 mm wide, *stiff*, greyish, wintergreen; ligules short, blunt. Inflorescence with 1 inconspicuous male spike above but overtopped by 2–4 *short* female spikes on *long, hair-like* stalks; female glumes yellowish. Fruit △b olive to dark brown, shortly beaked. Wet woods and hillsides, often on calcareous soils. B, S. Map 291. **2a** *C. krausei* is much shorter and has the male spike sometimes female at the tip and the female spikes more numerous and less bunched. Iceland only.

3. **Thin-spiked Wood Sedge** *Carex strigosa.* Loosely tufted; stems to 70 cm. Leaves *broad*, 6–10 mm wide, curved, green; ligules *pointed*. Inflorescence erect, with 1 male spike above 3–6 slender female spikes, whose stalks are *hidden* in the leaf-like bracts, which are shorter than the inflorescence; female glumes *green*. Fruit △c green, often curved, *very shortly* beaked. Wet places in woods, often on calcareous soils. B, F, G. Map 292.

4. **Starved Wood Sedge** *Carex depauperata.* Loosely tufted; stems to 100 cm. Leaves 2–4 mm wide, green to yellow-green; ligules short, blunt. Inflorescence with 1 male spike above 2–4 short female spikes, *well spread out* down the stem, their stalks *half-hidden* in the long leaf-like bracts, the upper of which sometimes exceed the inflorescence; female glumes brown. Fruit △d brownish-green, long-beaked. Dry, open woods and scrub, often on calcareous soils. B (rare), F, G. Map 293. **4a** *C. pilosa* is shorter (to 50 cm), with tufts of non-flowering stems and slender leafless flowering stems; leaves hairy along the edges and beneath; longer female spikes on longer stalks with red-brown glumes; and fruit △e shorter-beaked. F, G. Map 294.

The Sedge Family Cyperaceae

1. **Glaucous Sedge** *Carex flacca*.   One of the commonest sedges, especially in *calcareous* grassland. Loosely tufted, far creeping; stems to 60 cm, bluntly 3-sided. Leaves 2–6 mm wide, stiff, *glaucous beneath*, veined to tip. Inflorescence with 1–3 male spikes above 1–5 female spikes, the lowest bract about as long as the inflorescence; glumes purple-brown; flowering *April*–May. Fruit △a yellow-green to purple-black, scarcely beaked, rough. Chalk and limestone grassland, also dunes, fens and other damp calcareous places. T. Map 295.

2. **Carnation Sedge** *Carex panicea*.   Tufted, shortly creeping; stems to 50 cm, bluntly 3-sided. Leaves 2–4 mm wide, flat, *glaucous on both sides*, unveined at tip. Inflorescence with *a single male spike* above 1–3 female spikes, the lowest bract shorter than the inflorescence; glumes purple-brown. Fruit △b olive-green, often purple-tinged, scarcely beaked. Moors, mires and other *wet* places, not on very acid or very calcareous soils. T. Map 296. **2a.** **Sheathed Sedge** *C. vaginata* □ has yellowish- to apple-green leaves, the lowest bract often shorter than its spike and with a distinctly inflated sheath, and fruit △c asymmetrically beaked; on mountain ledges and in damp woods. B, S. Map 297. **2b** *C. livida* is shorter, with the grooved whitish-glaucous leaves longer than the inflorescence, and fruit △d more oblong and glaucous to yellow-green; bogs and other wet places. S. Map 298.

3. **Smooth-stalked Sedge** *Carex laevigata*.   Tufted; stems to 120 cm, sharply 3-sided. Leaves 6–12 mm wide, *bright to pale* green; ligules *long, pointed*. Inflorescence with 1–2 male spikes above 2–4 well-spaced female spikes, the lowest bract shorter than the inflorescence; glumes brown. Fruit △e green with reddish dots, or reddish-green, beaked. Damp, shady woods, often on clay soils. B, F, G. Map 299.

4. **Pale Sedge** *Carex pallescens*.   Tufted; stems to 60 cm, sharply 3-sided. Leaves 2–5 mm wide, pale to mid green, soft, often *hairy beneath*. Inflorescence short, with 1 male spike above and often hidden by 2–3 female spikes, *clustered near the top of the stem*, the lowest bract crimped and twisted at base and longer than the inflorescence; glumes pale brown. Fruit △f green, unbeaked. Open woods and damp grassland, often on clay soils. T. Map 300.

△ ×6

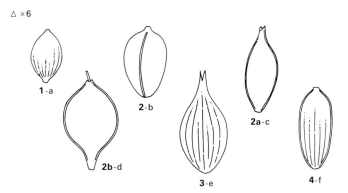

1-a

2-b

2a-c

2b-d

3-e

4-f

The Sedge Family Cyperaceae

1. **Green-ribbed Sedge** *Carex binervis.* Tufted, more loosely in drier places; stems to 150 cm, often much shorter. Leaves 3–6 mm wide, *dark green*, developing wine-red blotches. Inflorescence with 1 male spike above 2–3 female spikes, 15–45 mm long, well spaced down the stem; all bracts shorter than inflorescence; glumes purple-brown. Fruit △a *purple-brown* with two prominent *green ribs*. Moors, heaths and rough grassland on *acid soils.* T. Map 301.

2. **Distant Sedge** *Carex distans.* Tufted; stems to 100 cm. Leaves 2–6 mm wide, *grey-green*, fading to brown. Inflorescence with 1 male spike above 2–3 short female spikes, 10–30 mm long, compact at first, but well spaced out down the stem in fruit; all bracts usually shorter than inflorescence; glumes brown. Fruit △b *green*, not conspicuously ribbed. Damp places, especially sand or rocks *by the sea* and inland marshes. T. Map 302.

3. **Dotted Sedge** *Carex punctata.* Loosely tufted; stems to 100 cm. Leaves 2–5 mm wide, *pale green.* Inflorescence with 1 male spike above 2–4 short female spikes, 7–25 mm long, the lower spikes well spaced out down the stem; at least 1 bract usually *longer* than the inflorescence; glumes rufous-brown. Fruit △c pale green, *dotted red-brown*, markedly ribbed when dry, inserted at right angles on the inflorescence, giving a spiky appearance. Damp rocks and other bare or grassy places *by the sea*; rare inland. T. Map 303.

4. **Tawny Sedge** *Carex hostiana.* Loosely tufted; stems to 60 cm, slightly rough at the top. Leaves 2–5 mm wide, *3-sided* at the top, *yellowish-green.* Inflorescence with 1 male spike above 1–3 short female spikes, 7–20 mm long, spaced down the stem; bracts much shorter than inflorescence; glumes dark brown. Fruit △d *yellow-green*, ribbed. Marshes, fens, valley bogs, moorland flushes and other rather wet grassy places. T. Map 304.

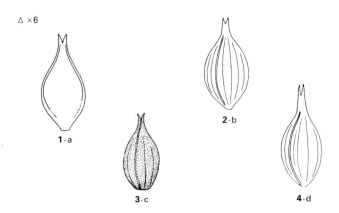

△ ×6

1-a

2-b

3-c

4-d

The Sedge Family Cyperaceae

△ ×6

1-a

1. **Long-bracted Sedge** *Carex extensa*. Tufted; stems to 40 cm, rather stiff. Leaves 2–3 mm wide, stiff, grooved, *grey-green*. Inflorescence with 1 male spike above 2–4 short un-stalked female spikes, usually clustered at the top, but the lowest sometimes stalked and well down the stem; bracts often spreading or *down-turned*, much *longer* than the inflorescence; glumes red-brown. Fruit △a *grey-green* or brown. *Saltmarshes* and bare ground by the sea. T. Map 305.

YELLOW SEDGES *Carex flava* agg. (2–5) A difficult group with several hybrids and intermediates. Tufted; stems 3-sided, smooth. Inflorescence with 1 male spike above 2–4 unstalked egg-shaped female spikes, clustered at the top, but the lowest usually stalked and well down the stem; the lowest bract usually longer than the inflorescence, often spreading or down-turned; flowering in May or June. Fruit usually abruptly narrowed to the beak, the lower usually down-turned when ripe. See Appendix 2 (p. 219).

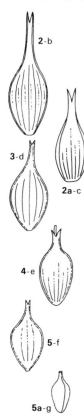

2-b

3-d

2a-c

4-e

5-f

5a-g

2. **Large Yellow Sedge** *Carex flava*. Stems to 70 cm. Leaves 4–7 mm wide, *broader* than other yellow sedges, bright yellow-green, often *longer* than the stems. Glumes orange-brown. Fruit △b *yellow*, larger than other yellow sedges, with a bent beak. Fens and marshes, usually on calcareous soils. T, but B (rare). **2a** *C. jemtlandica* has narrower, dark green leaves and smaller, more greenish fruit △c S. Map 306.

3. **Long-stalked Yellow Sedge** *Carex lepidocarpa*. Loosely tufted; stems to 75 cm, erect. Leaves 2–5 mm wide, up to half as long as stems, green or yellow-green, not wintergreen. Inflorescence with male spike always *stalked*; lowest bract often erect and sometimes only *equalling* inflorescence; glumes rufous-brown. Fruit △d yellow-green, with a bent beak. In Scotland may have shorter, broader leaves, dark brown glumes and dark green fruit. Fens, mountain flushes and other wet places on *calcareous* soils. T. Map 307.

4. **Common Yellow Sedge** *Carex demissa*. Distinctly tufted; stems to 40 cm, often *at an angle*. Leaves 2–3 mm wide, stiff, recurved, often *longer* than the stems, yellow-green, often wintergreen. Inflorescence with male spike *stalked*, male glumes orange-brown and female brown. Fruit △e yellow-green. Damp and wet places on only slightly calcareous or, more often, slightly *acid* soils. T. Map 308.

5. **Small-fruited Yellow Sedge** *Carex serotina*. Stems to 40 cm, erect or at an angle. Leaves 2–3 mm, *narrower* than other yellow sedges, stiff, usually flat, yellowish-green. Inflorescence with male spike usually *unstalked*, females usually all clustered; glumes orange- or yellow-brown. Fruit △f yellow-green, the lower usually *not down-turned*. Valley bogs, dune slacks, lake shores and other wet places on both slightly calcareous and slightly acid soils. T. Map 309. **5a** *C. bergrothii* has larger female spikes and the lower bract more erect. Fruit △g. S. Map 310.

Note: In these side-on drawings, the bending of the beaks in fruits 2-b and 3-d is *towards* the reader, and is thus not obvious in this diagrammatic form.

The Sedge Family Cyperaceae

△ ×6

1-a

1a-b

1b-c

2-d

3-e

3a-f

1: Fingered Sedge *Carex digitata.* Tufted; stems to 30 cm, leafless. Leaves 2–5 mm wide, usually sparsely hairy above, dark to yellowish-green, wintergreen, with purple basal sheaths. Inflorescence with 2–3 finger-like female spikes, *clustered* around and *overtopping* the thinner, scarcely stalked male spike; glumes red- or purple-brown, the male pointed; flowering *April*–May. Fruit △a yellowish-brown, *as long as* the glumes, abruptly narrowed to a short beak. Dry grassland and rocks, especially on calcareous soils. T. Map 311. **1a* Birdsfoot Sedge** *C. ornithopoda* □ is smaller in all its parts, with hairless pale green leaves, sometimes with reddish basal sheaths, unstalked male spike, blunt male glumes and fruit △b much longer than pale brown glumes; flowering May–June. Map 312. **1b** *C. pediformis* □ has leafy stems to 50 cm, narrower leaves, fatter female spikes, the lower spaced down the stem, and fruit △c gradually narrowed to the beak; flowering later. S. Map 313.

2: Dwarf Sedge *Carex humilis.* Tufted; stems to 10 cm. Leaves *thread-like*, 1–2 mm wide, often curved, pale green, becoming grooved, much *longer* than the stems. Inflorescence with 1 thin male spike above 2–4 slender female spikes *spaced* down the stem, their stalks hidden by brown sheaths; flowering *March*–April. Fruit △d greenish- or yellowish-brown. Dry grassland, especially on calcareous soils. B, F, G. Map 314.

3: Spring Sedge *Carex caryophyllea.* Loosely tufted; stems to 30 cm, almost leafless, usually rough at the top. Leaves *c.* 2 mm wide, *dark* green, wintergreen. Inflorescence with 1 male spike above 2–3 *clustered* female spikes, the lowest bract usually *shorter* than its spike, forming a sheath round the stem; glumes red-brown, the female with a short *green awn*; flowering *April*–May. Fruit △e green, downy, as long as the glume. Dry grassland, especially on *calcareous* soils, also in mountain flushes. T. Map 315. **3a** *C. hallerana* has more egg-shaped female spikes. F, G. Map 316. Fruit △f. **3b Shady Sedge** *C. umbrosa* has paler green leaves, the lowest bract usually as long as the spike, and female glumes shorter than the paler fruit △g; woods. F, G. Map 317. **3c** *C. glacialis* has narrower leaves, often curly; fruit △h; bare stony places. S. Map 318. 3a–3c are more tufted with female glumes unawned and shorter than the fruit, and flowering later.

4: Heath Sedge *Carex ericetorum.* Tufted; stems to 30 cm, almost leafless. Leaves 2–4 mm wide, dark green, pale-edged. Inflorescence with 1 male spike above 1–3 clustered female spikes, the lowest bract *very short*; glumes purple-brown with a *pale margin*, the female rounded and unawned; flowering *April*–May. Fruit △i green, downy, as long as the glume. Dry grassland and heaths, usually on calcareous soils. T. Map 319.

3b-g

3c-h

4-i

1b

1

1a

4

3

2

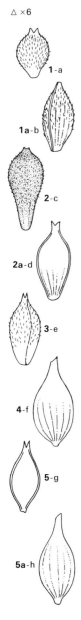

△ ×6

1-a

1a-b

1a-b

2-c

2a-d

3-e

4-f

5-g

5a-h

The Sedge Family Cyperaceae

1. **Downy-fruited Sedge** *Carex tomentosa*. Creeping; stems to 50 cm, rough at the top. Leaves *c.* 2 mm wide, *glaucous*. Inflorescence with 1–2 male spikes above 1–2 egg-shaped female spikes, mostly clustered, the lowest bract *longer* than its spike; glumes purple-brown. Fruit △a green, *downy*, not veined, with a very short, notched beak. Grassland, usually damp and *calcareous*. T. Map 320. **1a** *C. globularis* □ is loosely tufted, with 1 male spike and 2–3 female spikes, shortly spaced out, the lowest bract much longer than its spike; glumes dark brown, pale-edged; fruit △b distinctly veined; also in woods. S. Map 321.

2. **Soft-leaved Sedge** *Carex montana*. Makes large *matted tufts*; stems to 40 cm, lax, rough at the top, almost leafless. Leaves *c.* 2 mm wide, *soft*, somewhat *hairy* and pale green at first, turning golden, with basal sheaths *crimson*. Inflorescence with 1 male spike above 1–4 egg-shaped female spikes, all clustered, the lowest bract no longer than its spike; male glumes red-brown, female blackish, blunt. Fruit △c *blackish*, downy, unbeaked. Dry grassland, scrub and open woods, often on calcareous soils. T. Map 322. **2a** *C. fritschii* is generally larger in all its parts, with female glumes red-brown and pointed and fruit △d whitish, sometimes hairless and distinctly beaked. F (rare), G (rare). Map 323.

3. **Pill Sedge** *Carex pilulifera*. Tufted; stems to 30 cm, rough at the top, often curved or *decumbent*. Leaves *c.* 2 mm wide, green or yellowish-green. Inflorescence with 1 male spike above 2–4 egg-shaped female spikes, all clustered, the lowest bract *longer* than its spike; glumes red-brown, unawned. Fruit △e green, shortly downy. Dry grassland, heaths and moors on *acid* soils. T. Map 324.

4. **White-flowered Sedge** *Carex alba*. Loosely tufted, creeping, with distinctive white rhizomes; stems to 25 cm, usually *smooth* above. Leaves thread-like, *c.* 1 mm wide, *pale* green. Inflorescence with 1–3 short-stalked, clustered, egg-shaped female spikes, the upper usually overtopping the single male spike; glumes *whitish*, pointed. Fruit △f brown or blackish, veined, beaked. Dry stony hillsides, scrub and woods, mainly on calcareous soils. F, G.

5. **Small Sedge** *Carex supina*. Loosely tufted, creeping; stems to 20 cm, rough at the top. Leaves thread-like, *c.* 1 mm wide, *dark* green. Inflorescence with 1 male spike above 1–3 unstalked egg-shaped female spikes, sometimes *spaced* down the stem; female glumes red-brown, pointed. Fruit △g yellowish- to red-brown, shiny, *scarcely veined*, with a short *forked* beak. Dry grassy and heathy places, usually on sandy soils. G. Map 325. **5a Glossy-fruited Sedge** *C. liparocarpos* □ has broader leaves, 1–3 mm wide, the lowest female spike long-stalked, female glumes sometimes blunt and fruit △h distinctly veined with its beak not forked. F, G.

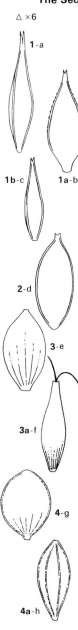

△ ×6

1-a

1b-c 1a-b

2-d

3-e

3a-f

4-g

4a-h

The Sedge Family Cyperaceae

1. Red-brown Sedge *Carex fuliginosa.* Tufted; stems to 30 cm, bluntly 3-sided. Leaves 2–5 mm wide. Inflorescence with 3–4 spikes, clustered, *thickened at tip*, the top one female above, *male at base*, the rest all female; glumes dark red- or blackish-brown. Fruit △a greenish-yellow to dark red-brown, *hairless*, not veined, gradually narrowed to the beak. Stony and rocky places on *mountains*, often on calcareous soils. S. Map 326. **1a Evergreen Sedge** *C. sempervirens* forms tussocks and has longer, laxer leaves, egg-shaped spikes, the top one all male, and fruit △b obscurely veined and sometimes downy; also in grassy places. F, G. Map 327. **1b Short-spiked Sedge** *C. brachystachys* is creeping, with narrower leaves, narrowly cylindrical spikes, the top one sometimes all male, and has paler fruit △c. F, G. Map 328.

2* Scorched Alpine Sedge *Carex atrofusca.* Loosely tufted, creeping; stems to 35 cm. Leaves 2–5 mm wide, mid-green, soft, with a 3-sided tip. Inflorescence with 3–5 egg-shaped spikes, the top one either all male or *partly female*, the lowest one sometimes spaced down the stem; glumes blackish-red or *blackish-purple*. Fruit △d *purple-black*, hairless, not veined, abruptly narrowed to the beak. Wet stony places on *mountains*. B, S. Map 329.

3* Bog Sedge *Carex limosa.* Loosely tufted, creeping; stems to 40 cm, *sharply* 3-sided, rough, stiff. Leaves *c.* 1 mm wide, rough, grooved, *glaucous*. Inflorescence with 1 thin male spike above 1–3 oblong female spikes, *drooping* on long stalks, the lowest bract *shorter* than the inflorescence; glumes red- or purple-brown, pointed. Fruit △e *glaucous*, veined, abruptly narrowed to a very short beak. Wet bogs, bog pools and peaty lake shores, mainly at *lower altitudes*. T. Map 330. **3a** *C. laxa* has rather lax, bluntly 3-sided smooth stems, flat leaves, blunter female glumes, and fruit △f more gradually narrowed to the beak. S. Map 331.

4* Tall Bog Sedge *Carex magellanica.* Shortly creeping; stems to 40 cm, *bluntly* 3-sided, smooth. Leaves 2–3 mm wide, pale green, often as long as stems. Inflorescence with 1 thin male spike above 2–4 long-stalked egg-shaped female spikes, the top one often *female at the tip* and the lower ones with a few *male flowers at the base*, the lowest bract usually *longer* than the inflorescence; glumes red- or purple-brown. Fruit △g blue-green, obscurely veined, *broader* than its glume. Wet bogs. B, S. Map 332. **4a* Mountain Bog Sedge** *C. rariflora* has shorter stems, narrower leaves shorter than the stems, top spike all male and shorter-stalked lower ones all female, the lowest bract shorter than the inflorescence, darker glumes and fruit △h markedly veined, obscurely beaked, and not broader than its glume; peaty places on mountains, and on tundra. Map 333.

The Sedge Family Cyperaceae

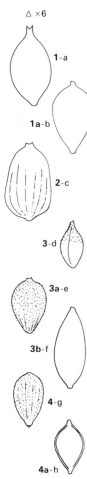

△ ×6

1-a

1a-b

2-c

3-d

3a-e

3b-f

4-g

4a-h

1. **Black Alpine Sedge** *Carex atrata*. Tufted; stems medium, to 60 cm, sharply *3-sided*. Leaves 3–10 mm wide, pale green or *glaucous*, no longer than stems. Inflorescence with 3–5 stalked *nodding* egg-shaped spikes, clustered near the top, the lowest usually longer-stalked and down the stem, the top one female above and *male at the base*; lowest bract leaflike, longer than the inflorescence; glumes *purple-black*. Fruit △a greenish- or brownish-yellow, often marked dark purple, not veined, gradually narrowed to the beak. Grassy and rocky places, usually in *mountains*. B, S. Map 334. **1a Small-flowered Sedge** *C. parviflora* □ is shorter, with narrower (2–4 mm) leaves, sometimes longer than stems, smaller flower spikes, not or scarcely stalked, the top spike sometimes all male or male at the top, and the lowest bract usually spreading or down-turned. Fruit △b. F, G.

2. **Club Sedge** *Carex buxbaumii*. Loosely tufted, creeping; stems medium tall, to 80 cm, sharply 3-sided, roughish, stiff. Leaves 2–4 mm wide, *glaucous*, the margins inrolled. Inflorescence with 2–5 erect spikes, the top one *club-shaped*, female above, *male at the base*, the lower egg-shaped, unstalked and *spaced down* the stem; lowest bract as long as or longer than the inflorescence; glumes dark red-brown. Fruit △c *green*, faintly *ribbed*, very shortly beaked. Fens and wet grassland. T, but B (rare). Map 335. **2a** *C. hartmanii* □ has darker green leaves, inflorescence nodding, all spikes cylindrical and usually overlapping each other, and a shorter lower bract. G, S.

3. **Close-headed Alpine Sedge** *Carex norvegica*. Tufted; stems medium, to 60 cm, sharply 3-sided. Leaves 2–5 mm wide, mid-green. Inflorescence with 2–4 unstalked *erect* egg-shaped spikes, clustered at the top, the top one female above and *male at the base*; lowest bract leaf-like, as long as or longer than the inflorescence; glumes *purplish-black*. Fruit △d greenish- to greyish-brown, blackish at tip, not veined, shortly beaked. Damp places, stony, grassy or wooded. B, S. Map 336. **3a** *C. holostoma* is scarcely tufted, creeping and with shorter roughish stems; shorter, narrower (*c.* 2 mm) yellowish- or greyish-green leaves; the top spike all male and overtopped by the females; darker glumes and scarcely beaked fruit △e. S. Map 337. **3b** *C. stylosa* has roughish stems, narrower (2–3 mm), sometimes greyish-green leaves, the top spike all male, all spikes shortly stalked, and fruit △f gradually narrowed to a longer beak. S.

4. **Two-coloured Sedge** *Carex bicolor*. Loosely tufted; stems short, to 20 cm, sometimes *prostrate*, sharply 3-sided. Leaves 1–2 mm wide, greyish- or yellowish-green. Inflorescence with 2–4 nodding unstalked egg-shaped clustered spikes, the top one female above and *male at the base*, the lowest short-stalked; lowest bract as long as or longer than inflorescence; glumes dark red- to blackish-brown; styles 2. Fruit △g pale greyish-green, not beaked or veined, *dotted*. Damp bare places. S. Map 338. **4a** *C. rufina* has stems curved, shorter than the paler green leaves, and fruit △h very shortly beaked and not dotted; snow patches. Map 339.

The Sedge Family Cyperaceae

△ ×6

1-a

1a-b

2-c

3-d

3a-e

1. Northern Estuarine Sedge *Carex paleacea*. Creeping; stems to 50 cm, sharply 3-sided, their faces *concave*. Leaves 4–8 mm wide, yellowish-green, as long as or longer than the stems. Inflorescence with 1–3 male spikes above 5–7 female spikes, *spaced down the stem*, stalked and drooping; the lowest bract longer than the inflorescence; female glumes yellowish- or reddish-brown, *long-awned*; styles 2. Fruit △a greenish, faintly veined, shortly beaked. Estuaries and saltmarshes. S. Map 340. **1a** *C. lyngbyei* has shorter leaves and darker female glumes with no or only a short awn. Fruit △b. Map 341. **1b** *C. vacillans* is intermediate between Northern Estuarine and Common Sedges (p. 164) and probably their hybrid. Map 342.

2. **Estuarine Sedge** *Carex recta*. Probably a partially fertile hybrid of Northern Estuarine and Water Sedges (1 and 4). Creeping, with tufted shoots and sometimes tussocks; stems to 100 cm, their faces *flat*. Leaves 3–6 mm wide, rough, mid- to yellowish-green, longer than the stems. Inflorescence with 1–4 male spikes above 2–4 *overlapping* female spikes, the upper often male at the top, the lowest bract longer than the inflorescence; female glumes dark red-brown with the vein projecting from the tip; styles 2. Fruit △c green, faintly veined, shortly beaked. Muddy estuaries in north-east Scotland. Map 343. **2a** *C. halophila* is a very similar putative hybrid in north Scandinavia. Map 344.

3. Arctic Saltmarsh Sedge *Carex subspathacea*. Far-creeping; stems *very short*, to 5 cm, bluntly 3-sided. Leaves *c.* 4 mm wide, grey-green, as long as or longer than the stems. Inflorescence with 1 male spike above 2–3 overlapping short- or unstalked female spikes, the lowest bract as long as the inflorescence; female glumes *dark* or reddish-brown; styles 2. Fruit △d grey-green, scarcely veined, gradually narrowed to the beak. By the sea, mainly on mud and gravel. S. Map 345. **3a** *C. salina* is probably the hybrid of Northern Estuarine Sedge (1) with Arctic Saltmarsh Sedge; stems to 20 cm, leaves narrower but longer than the stems; lowest bract sometimes longer than the inflorescence, paler and more pointed female glumes, and fruit △e pale yellowish- or greenish-brown. Map 346.

4. **Water Sedge** *Carex aquatilis*. Loosely tufted, creeping; stems to 110 cm, brittle, bluntly 3-sided. Leaves 3–5 mm wide, glaucous above, *bright green beneath*. Inflorescence with 2–4 male spikes above 2–5 overlapping stalked female spikes, the upper often male at the top, the lowest often down the stem; the lowest bract longer than the inflorescence; female glumes reddish- or purplish-brown, *blunt*; styles 2. Fruit △f green, unveined, *unbeaked*. Swamps, riversides, lake shores. Also on mountains, where smaller, with pale green leaves, 1 male spike only, female glumes pointed and fruit brown; this is var. *minor* or perhaps the hybrid with Stiff Sedge (p. 164). B, G, S. Map 347.

4-f

The Sedge Family Cyperaceae

△ ×6

1-a

2-b

2a-c

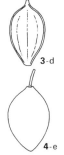

3-d

4-e

5-f

1. **Stiff Sedge** *Carex bigelowii.* Creeping; stems to 30 cm, but often much shorter, *stiff*, sharply 3-sided. Leaves 2–7 mm, often *curved*, glaucous. Inflorescence with 1 male spike above 2–3 unstalked female spikes, clustered together, or the lowest down the stem, with the lowest bract quite *short*; female glumes purple-black. Fruit △a green, usually partly purple-black, shortly beaked, with *2 styles. Mountains*, often common on high-level heaths and bare stony places. B, G, S. Map 348.

2. **Tufted Sedge** *Carex elata.* Forming *tussocks* up to 40 cm high; stems tall, to 100 cm, rough, sharply 3-sided. Leaves 3–6 mm wide, rough, *glaucous*; basal sheaths fibrous. Inflorescence with 1–3 male spikes (the lowest sometimes female at base) above 2–3 unstalked female spikes, fairly close together, with the lowest bract much *shorter* than the inflorescence; female glumes blackish-brown, bluntly pointed. Fruit △b red-brown, often partly red-brown, unveined, shortly beaked, with *2 styles, soon falling.* Fens, swamps and shallow still fresh water. T. Map 349. **2a Lesser Tufted Sedge** *C. cespitosa* is shorter and slenderer, with narrow bright or yellowish-green leaves, only 1–2 female spikes and fruit △c grey- or brownish-green. Wet meadows, alder carr. F, G, S. Map 350.

3. **Common Sedge** *Carex nigra.* A very variable sedge, creeping, tufted or (var. *juncea*) forming tussocks; stems tall, to 70 cm, but often much shorter, rough, *bluntly* 3-sided. Leaves 2–3 mm, glaucous. Inflorescence with 1–2 male spikes above 1–4 unstalked female spikes, fairly close together, or the lowest spike stalked and down the stem; the lowest bract shorter than or about *as long as* the inflorescence; female glumes *blackish*, bluntly pointed. Fruit △d green, usually partly blackish, shortly beaked, with *2 styles.* Mires, marshes, grassy moorland, dune slacks and other wet or silty places. T. Map 351.

4. **Slender Tufted Sedge** *Carex acuta.* Tufted and creeping; stems tall, to 120 cm, rough and sharply 3-sided. Leaves 3–7 mm wide, *glaucous*, rough-edged, often drooping at the tips. Inflorescence with 1–3 male spikes (the lowest often female at base) above 2–4 mostly unstalked female spikes, fairly close together, the lowest bract *longer* than the inflorescence; female glumes blackish- or red-brown, usually *pointed.* Fruit △e, *veined, not* beaked, with *2 styles* (cf. Lesser Pond Sedge, p. 142). Marshes, freshwater margins and other wet places. T. Map 352.

5. **Three-nerved Sedge** *Carex trinervis.* Creeping; stems short, to 30 cm, *bluntly* 3-sided. Leaves *c.* 2 mm wide, greyish, *grooved.* Inflorescence with 1–4 male spikes above 2–4 female spikes, close together, the lowest bract *longer* than the inflorescence; female glumes brownish, shortly *awned.* Fruit △f yellowish- to grey-green, often spotted purple; styles 2. Damp sandy and heathy places *near the sea.* F, G. Map 353.

The Sedge Family Cyperaceae

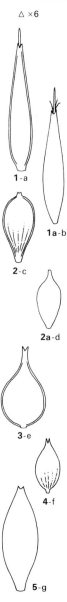

△ ×6

1-a

1a-b

2-c

2a-d

3-e

4-f

5-g

1.* Few-flowered Sedge *Carex pauciflora.* Loosely tufted, creeping and mat-forming; stems short, to 25 cm, slender, bluntly *3-sided.* Leaves thread-like, *c.* 1 mm wide. Inflorescence with a single *bractless* spike, male above, 2–3 female flowers at base; glumes pale red-brown. Fruit △a *straw-coloured, down-turned.* Bogs and wet peaty moors. T. Map 354. **1a.* Bristle Sedge** *C. microglochin* is shorter, with stems almost rounded and narrower, slightly grooved leaves, 3–12 female flowers and a bristle projecting from the tip of the fruit △b. Base-rich flushes in mountains. B, S. Map 355.

2. Rock Sedge *Carex rupestris.* Tufted and creeping; stems short, to 20 cm, bluntly 3-sided. Leaves *c.* 1 mm wide, *curly.* Inflorescence with a single spike, male above and 3–6 female flowers at base, with a short *bristle-like* bract; glumes red- or purple-brown with a pale margin, as long as or longer than fruit; a shy flowerer. Fruit △c *greenish-brown,* matt. Dry rock ledges or stony slopes on base-rich soils in *mountains.* B, S. Map 356. **2a** *C. obtusata* is not tufted and has straight leaves, 5–10 female flowers, their glumes with a broader pale margin, much shorter than the shiny red-brown fruit △d. Dry sandy places. G, S. Map 357.

3. Capitate Sedge *Carex capitata.* Tufted; stems short to medium, to 40 cm, bluntly *3-sided, rough* at top. Leaves *bristle-like,* usually *shorter* than stems. Inflorescence with a single *egg-shaped* spike, male above and female at base; glumes yellow- to red-brown. Fruit △e greenish, often partly brown, *unveined.* Damp, usually base-rich and peaty ground. S Map 358.

4. Cushion Sedge *Carex nardina.* Forms flattened *cushions;* stems low, to 10 cm, *curved,* sometimes half-prostrate, *rounded, smooth.* Leaves bristle-like, usually *longer* than stems, usually curved and often *prostrate.* Inflorescence with a single spike, male above and female at base; glumes red-brown. Fruit △f pale brown, veined. Dry, stony, calcareous places in mountains. S. Map 359.

5.* Flea Sedge *Carex pulicaris.* Tufted, creeping; stems medium, to 30 cm, *rounded.* Leaves thread-like, *c.* 1 mm wide, dark green. Inflorescence with a single spike, male at top, female at base; glumes red- or purple-brown. Fruit △g dark brown, shining, *down-turned* when ripe and female glumes have fallen off. *Fens,* flushes and other damp places irrigated by calcareous water. T. Map 360.

The Rush Family Juncaceae

RUSHES *Juncus* (pp. 168–178) are hairless perennials (a few are annual), usually erect and tufted, growing in damp or wet places. Leaves, when present, narrow. Unlike grasses and sedges, rushes have recognisable small green, brown or yellowish flowers with 6 tepals △a (i.e. no distinction between petals and sepals), appearing in late June and July. Fruit a small 3-valved capsule, often important in identification.

△ ×4

a

1. **Sea Rush** *Juncus maritimus*. Loosely tufted, *creeping*; stems tall, to 100 cm, *c.* 2 mm wide, stiff, pale grey-green. Leaves at base, *c.* 2 mm wide, rounded, *sharply* pointed. Inflorescence a loose forking cluster below a long, *sharply* pointed bract, appearing as a continuation of the stem; flowers pale *yellow*. Fruit △b brown, bluntly pointed, *as long as* or slightly longer than the tepals. Drier *saltmarshes* and among rocks by the sea, also in brackish meadows inland. T. Map 361.

1-b

2. **Sharp Rush** *Juncus acutus*. Forming *substantial clumps*; stems tall to *very tall*, to 150 cm, rigid. Leaves at base 2–4 mm wide, ending in a long exceedingly *sharp spine*. Inflorescence a fairly tight forking cluster below a long *very sharp* spine, appearing as a continuation of the stem; flowers *red-brown*. Fruit △c brown, almost globular, much *longer* than the tepals. *Dunes* and dune slacks by the sea; rare in saline places inland. B, F Map 362.

2-c

3. **Fine-leaved Rush** *Juncus subulatus*. Patch-forming; stems medium to tall, to 120 cm. Leaves narrow, *all on the stems*. Inflorescence a forking cluster with the lowest bract much *shorter* than the inflorescence; flowers pale yellow-green to brownish. Fruit △d red-brown, glossy, 3-sided, somewhat *shorter* than the tepals. Only in Britain, in one dryish saltmarsh in Somerset.

3-d

4. **Round-fruited Rush** *Juncus compressus*. Loosely tufted, creeping; stems short to medium, to 40 cm, usually *flattened*. Leaves *c.* 1 mm wide, usually flat, glaucous, both on stems and at base. Inflorescence a loose forking cluster, the lowest bract usually *longer* than the inflorescence; flowers pale brown. Fruit △e glossy dark brown, bluntly egg-shaped, as long as or *longer* than the tepals. Fens, marshes and damp grassland, often with bare patches, usually on base-rich *non-saline* soils. T. Map 363.

4-e

5. **Saltmarsh Rush** *Juncus gerardi*. Creeping and making *extensive patches*; stems low to medium, to 50 cm, sometimes flattened. Leaves at base and usually also on stems, *c.* 1 mm wide, flat to roundish, dark green. Inflorescence a loose spreading cluster, the lowest bract usually shorter than the inflorescence; flowers dark brown. Fruit △f pale or chestnut brown, egg-shaped, about *as long as* the tepals. Salt- and brackish marshes, usually *by the sea*; the commonest small maritime rush. T. Map 364.

5-f

The Rush Family Juncaceae

△ ×4

1-a

2-b

1. **Hard Rush** *Juncus inflexus.* Tufted; stems medium to tall, to 120 cm, *hard,* greyish, *ridged, leafless,* containing *interrupted* pith. Inflorescence a loose cluster, with unequal stalks, below a long bract appearing as a continuation of the stem; flowers brown; flowering May–July. Fruit △a brown, egg-shaped, minutely pointed. Damp grassy places, liking lime and heavy soils; common. T. Map 365.

2. **Baltic Rush** *Juncus balticus.* Creeping, the stems *in rows,* medium to tall, to 100 cm, *greyish,* smooth and with *continuous* pith. Inflorescence a loose cluster, the terminal bract much *shorter* than in Hard Rush (1), and flowers darker brown. Fruit △b blunter than Hard Rush's and darker brown. Dune slacks and other damp sandy places. B, G, S. Map 366. **2a Arctic Rush** *J. arcticus* is much shorter, to 40 cm, with a still shorter bract and anthers shorter than filaments (longer in Baltic). Hybridises with Thread Rush (3). Damp places in mountains. S. Map 367.

3-c

3. **Thread Rush** *Juncus filiformis.* Creeping, with stems *in rows,* like Baltic Rush (2), but has *green,* markedly *slender,* slightly ridged stems, to 60 cm, a much longer bract and fewer, greenish or yellowish flowers. Fruit △c almost globular. Lakesides, poor fens and other damp or wet grassy places. T. Map 368.

4-d

4. **Soft Rush** *Juncus effusus.* Tufted; stems medium to very tall, to 150 cm, green, leafless, either *smooth* or faintly ridged, and with *continuous* pith. Inflorescence rather small, in a usually loose but sometimes fairly tight cluster, below a long bract appearing as a continuation of the stem; flowers greenish-brown; flowering late May to July. Fruit △d brown, egg-shaped; indented at the top, usually *shorter* than the tepals. Damp, often grassy places, especially in overgrazed, badly drained fields; common. T. Map 369.

5-e

5. **Compact Rush** *Juncus conglomeratus.* Differs from the tight-clustered form of Soft Rush (4) in having stems only to 100 cm and *conspicuously ridged;* darker brown flowers, sometimes in a small group of tight clusters but occasionally with a lax inflorescence; and fruit △e *as long as* the tepals; also in *avoiding limy* soils. T. Map 370.

The Rush Family Juncaceae

△ ×6

1-a

2-b

2a-c

3-d

4-e

1.⁕ Three-leaved Rush *Juncus trifidus.* Tufted, creeping and mat-forming; stems low to medium, to 40 cm, slender. Leaves *c.* 1 mm wide, *very short*, at base of stems. Inflorescence with 1–3 small tight clusters of 1–3 dark brown flowers and 2–3 *long leaf-like bracts* at base of each cluster. Fruit △a egg-shaped, shortly beaked, longer than the tepals. Bare, rocky and grassy places on *mountains*. B, S. Map 371.

2. Sand Rush *Juncus tenageia.* Tufted *annual*; stems low to short, to 30 cm. Leaves short, thread-like, including 3 on stem. Inflorescence a very open forking cluster, starting *more than half-way* up the stem, with single flowers at the base of the forks and up the stem, and *no* conspicuous bracts; flowers *brown*, the inner tepals *blunt*, the outer pointed. Fruit △b brown, *globular*, equalling tepals. Damp, rather bare places, exposed mud and tracks, usually on *acid* soils. F, G. Map 372. **2a** *J. sphaerocarpus* ☐ may have curved stems, with 0–2 stem leaves, the inflorescence usually starting less than half-way up the stem, long lower bracts, pointed inner tepals and egg-shaped fruit △c, much shorter than tepals. Map 373.

3.⁕ Toad Rush *Juncus bufonius.* Loosely tufted *annual*; stems low to medium, to 50 cm, lax. Leaves *c. 2 mm* wide, grooved, numerous, at base and on stems. Inflorescence a spreading, forking cluster, usually starting at least *half-way* up the stem, with single flowers in the forks and up the stem, and *1–5* leaf-like lower bracts, long but shorter than the inflorescence; flowers *greenish-white* with pointed tepals, the outer with a *narrow* pale margin, and anthers *no* longer than filaments. Fruit △d brown, egg-shaped, blunt or pointed, shorter than tepals. Wet mud and other bare damp places; common. T. Map 374. **3a⁕ Leafy Toad Rush** *J. foliosus* often has broader leaves and the inflorescence starting near the base of the stem, tepals marked with dark brown, anthers much longer than filaments, and fruit about as long as tepals. B, F. **3b⁕ Dwarf Toad Rush** *J. minutulus* is more tufted and much shorter, to 5 cm, with narrower leaves, 0–1 lower bracts and shorter anthers; usually on sandy soil. **3c⁕ Frog Rush** *J. ranarius* is shorter, to 20 cm, with narrower leaves, 1–4 lower bracts, some flowers usually in small heads, outer tepals with a broad pale margin, and fruit as long as or slightly longer than tepals; often in saltmarshes and grassy places. 3, 3a, 3b and 3c are mapped together, as their relative distribution is not well known.

4.⁕ Bulbous Rush *Juncus bulbosus* (incl. *J. kochii*). Very variable, tufted or *floating*; stems low to short on land, to 30 cm, but up to 100 cm in water, with a *bulbous* base. Leaves hollow, grass-like or bristle-like, rounded, slightly flattened or grooved, often jointed, at base or on stems; often bronzy or reddish when in water. Inflorescence a forking cluster of up to 20 heads, each with up to 15 green or brown flowers, often replaced by *green shoots*. Fruit △e brown, 3-sided, blunt, as long as or longer than tepals. Bogs, *wet* heaths and wet cart ruts, on *acid* soils. T. Map 375.

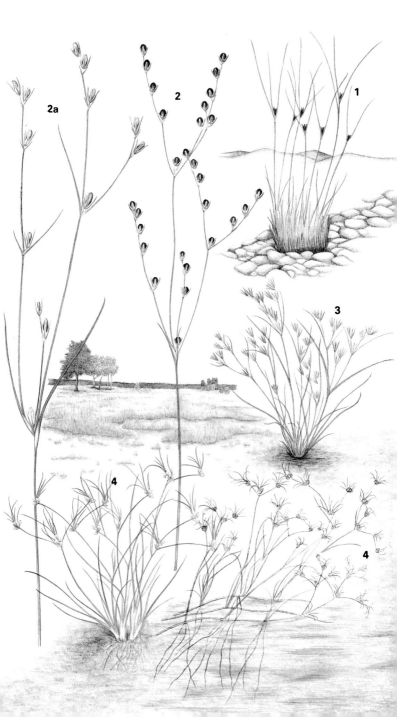

The Rush Family Juncaceae

1.* Heath Rush *Juncus squarrosus*. Tufted or mat-forming; stems short to medium, to 50 cm, *stiff*, usually leafless. Leaves 1–2 mm wide, wiry, grooved, in a *basal rosette*. Inflorescence a cluster of tight heads on unequal stalks, the lowest bract *shorter* than the inflorescence; flowers dark brown with conspicuously *pale margins*, giving them a silvery look; flowering late May to July. Fruit △a brown, egg-shaped, shorter than tepals. Heaths and moors on *acid* soils, especially with Mat-grass (p. 106); common. T. Map 376.

2.* Slender Rush *Juncus tenuis*. Tufted; stems short to tall, to 80 cm. Leaves 1–2 mm wide, flat or grooved, pale green, at base of and as long as the stems; long *whitish auricles* at base. Inflorescence a cluster of small beads on unequal stalks, the 2 lowest bracts usually *much longer* than the inflorescence; flowers greenish, becoming yellowish, with narrow, acutely pointed tepals. Fruit △b grey-brown, egg-shaped, shorter than tepals. Bare damp sandy places, especially tracksides, on *acid* soils. (T). Map 377. **2a* Dudley's Slender Rush** *J. dudleyi*, with leaves less than half as long as stem and yellowish auricles, is a North American plant naturalised in 2 places in Scotland only.

3.* Grass-leaved Rush *Juncus planifolius*. A recent arrival in *Ireland*, liable to be confused with Field Wood-rush (p. 180), but readily distinguished by its hairless leaves. Tufted; stems short, to 30 cm, leafless. Leaves *2–5 mm* wide, flat, pale green, shining, *grass-like*. Inflorescence a cluster of *tight* stalked heads, each with 8–10 dark brown flowers, the lower bract *short* and leaf-like. Fruit △c chestnut-brown, glossy, 3-sided, longer than tepals. Bare damp places, often by tracks and lakes. Native of Australia and the Pacific. (B).

△ ×6

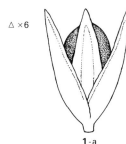

1-a

2-b

3-c

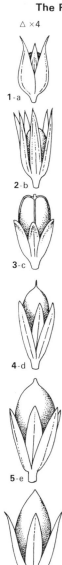

△ ×4

1-a

2-b

3-c

4-d

5-e

5a-f

1.* Dwarf Rush *Juncus capitatus.* *Annual;* stems low to short, 1–20 cm, *leafless.* Basal leaves *c.* 1 mm wide, wiry, grooved, soon withering. Inflorescence a tight cluster of 1–4 heads, each with 5–10 (exceptionally only 1) greenish flowers, which turn red-brown, the inner pair *shorter and colourless;* 2 bracts, 1 longer than the inflorescence; flowering May–June. Fruit △a pale brown, egg-shaped, markedly *shorter* than tepals, whose points are *curved downwards.* Damp, often bare places on heaths and in fens; decreasing. T, but B (rare). Map 378.

2.* Pigmy Rush *Juncus pygmaeus.* Tufted *annual;* stems low, 1–10 cm, sometimes with 1 leaf. Basal leaves less than 1 mm wide, thread-like, with pale auricles at base. Inflorescence a tight cluster of 1–5 heads, each with 2–15 green flowers; bracts both *shorter* than the inflorescence, distinguishing it at once from small Toad Rushes (p. 172), with which it often grows; flowering June. Fruit △b pale brown, egg-shaped, markedly *shorter* than the *very narrow spreading* tepals. Damp sandy and peaty places, especially in tracks; decreasing. B (rare), F, G. Map 379.

3.* Two-flowered Rush *Juncus biglumis.* Tufted, creeping; stems low to short, 2–20 cm, *leafless.* Leaves thread-like. Inflorescence a *small head* of usually only 2 dark brown flowers, one above the other, with a *bract longer* than the inflorescence. Fruit △c brown, very blunt, *longer* than tepals. Bare wet places on base-rich soils on *mountains.* B, S. Map 380.

4.* Three-flowered Rush *Juncus triglumis.* Differs from Two-flowered Rush (3) in having sometimes *1 stem-leaf,* usually *3* paler brown or yellowish flowers in the head, bracts *shorter* than the inflorescence, and more pointed fruits. Fruit △d. Wet places on more *acid* soils on *mountains.* B, S. Map 381.

5.* Chestnut Rush *Juncus castaneus.* Creeping; stems short, to 30 cm, sometimes with 1 leaf. Basal leaves *4 mm* wide, flat or grooved. Inflorescence a *tight cluster* of 1–3 heads, each with 3–20 chestnut to blackish-brown flowers, the lower bract *longer* than the inflorescence. Fruit △e red-brown, glossy, egg-shaped, *longer* than tepals. Wet bare and grassy places on *mountains* and lower down in the north. B, S. Map 382. **5a Bog Rush** *J. stygius* □ has 1–3 stem-leaves, stems and leaves often reddish, 1–2 heads of 2–3 flowers arranged in a row, flowers green to pale brown and fruit △f pale brown to yellowish; bogs on acid soils. G, S. Map 383.

The Rush Family Juncaceae

All rushes on this page have *jointed* leaves, on the *stems* only, the inflorescence in at least 2 *whorls* of loose forking clusters of small heads, the lowest bract quite *short*, and fruit longer than tepals. (See also p. 10.)

△ ×6

1-a

1a-b

2-c

2a-d

3-e

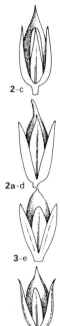

4-f

1. **Blunt-flowered Rush** *Juncus subnodulosus*. Creeping; stems medium to tall, to 100 cm. Leaves 2–4 mm wide, rounded. Inflorescence with 5–10 flowers in each head, *yellow*-brown at first, becoming red-brown, the tepals blunt. Fruit △a pale brown, 3-sided, *abruptly* narrowed to a short beak. Fens, marshes and other wet places, usually on *base-rich* soils. T. Map 384. **1a* Canadian Rush** *J. canadensis* □ a North American plant naturalised in Belgium and the Netherlands, is tufted and has green flowers with pointed tepals and fruit △b only minutely beaked; usually on acid soils. (F, G). **1b Sword-leaved Rush** *J. ensifolius*, also North American, naturalised only in Finland, has flattened stems, iris-like leaves □ and a shorter fruit.

2. **Sharp-flowered Rush** *Juncus acutiflorus*. Creeping; stems medium to tall, to 100 cm, *erect*. Leaves 2–4 mm wide, *rounded*. Inflorescence in 2 or more whorls, each head with 5–8 greenish-brown flowers, the tepals sharply *pointed*. Fruit △c chestnut-brown, *gradually* contracted to a sharp *point*. Hybridises with Jointed Rush (4). Marshes and wet grassy places, usually on *acid* soils; common. T. Map 385. **2a Black Rush** *J. atratus* has angled leaves, blackish-brown flowers and the fruit △d abruptly contracted to a long beak. G. Map 386.

3. **Alpine Rush** *Juncus alpinus*. Creeping; stems *low* to medium, to 60 cm. Leaves 2–3 mm wide, rounded. Inflorescence has each head with 6–8 dark red-brown flowers, the tepals *blunt* (*J.a.* ssp. *nodulosus* has flowers greenish to pale brown and tepals more pointed). Fruit △e brown, *blunt*, minutely pointed, sometimes no longer than tepals. Hybridises with Jointed Rush (4). Wet, usually rather bare places, mainly in *mountains*, but also very locally in lowlands. B, G, S. Map 387.

4. **Jointed Rush** *Juncus articulatus*. Tufted or creeping; stems low to medium, to 60 cm. Leaves 2–3 mm wide, *curved, flattened*, often *half-prostrate*. Inflorescence has each head with 5–15 brown flowers, the outer tepals usually pointed, the inner often *blunt*; appears a fortnight later than Sharp-flowered Rush (2). Fruit △f brown, egg-shaped, *abruptly* narrowed to a small beak. Hybridises with Sharp-flowered and Alpine Rushes (2 and 3) and *J. anceps*. Marshes, wet moors and grassland, usually on *acid* soils; common. T. Map 388. **4a** *J. anceps* □ has grooved leaves, the inflorescence often clearly divided into two parts and all tepals more or less blunt; hybridises with Jointed Rush. F, G, S. Map 389.

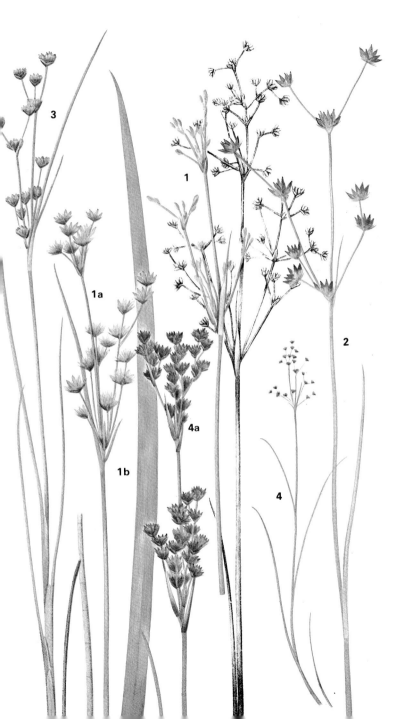

3

1

1a

2

1b

4a

4

The Rush Family Juncaceae

WOOD-RUSHES *Luzula* (pp. 180—182) are tufted perennials, easily recognised by the long *white hairs* that fringe their flat grass-like leaves, which are both basal and on the stems. Inflorescence a terminal cluster of rush-like flowers, with 6 tepals. Growing in drier places than rushes.

△ ×6

1. **Field Wood-rush** *Luzula campestris*. Loosely tufted, with *short* runners; stems low to short, to 30 cm. Leaves 2—4 mm wide. Inflorescence a cluster of 1 unstalked and 3—8 stalked heads of 5—12 chestnut-brown flowers, with conspicuous *bright yellow* anthers, much *longer* than filaments; lower bract short; flowering April—May. Fruit △a brown, globular, shorter than tepals. Dry grassland, especially on *basic* or neutral soils. T. Map 390. **1a** *L. sudetica* has narrower leaves, all heads stalked, unequal purple-brown tepals and anthers as long as filaments △b; wet places, mainly on acid peaty soils. G, S. Map 391.

1-a

2. **Heath Wood-rush** *Luzula multiflora*. Tufted, with *no* runners; stems short, to 30 cm. Inflorescence a cluster of 3—10 stalked heads, each with 8—18 red-brown flowers, anthers about *as long as* filaments and lower bract short; flowering May—June. Fruit △c brown, globular, shorter than tepals T. Map 392. **2a** *L.m.* ssp. *congesta* □ has unstalked heads, lower bract longer than inflorescence and fruit as long as tepals. Heaths, moors, grassland and woods on *acid* soils throughout the region. **2b*** **Fen Wood-rush** *L. pallescens* □ has pale or yellow-green leaves, the central cluster scarcely stalked, and flowers pale yellow-brown; not especially on acid soils. B (rare), G, S. Map 393.

1a-b

3. **Curved Wood-rush** *Luzula arcuata* (incl. *L. confusa*). Tufted, with short runners; stems low to short, to 25 cm. Leaves 1—3 mm, stiff, *grooved*, usually *hairless*. Inflorescence a cluster of 2—8 heads of 3—10 chestnut-brown flowers, some on short erect and others on longer *curved* stalks, the lower bract quite short; flowering June—July. Fruit △d brown, *shorter* than tepals. Wet stony places on *mountains* and tundra. B, S. Maps 394 and 395. **3a** **Arctic Wood-rush** *L. arctica* □ has flat leaves, unstalked clusters and fruit △e slightly longer than tepals; on calcareous soils. S. Map 396.

2-c

4. **Spiked Wood-rush** *Luzula spicata*. Tufted; stems low to short, to 25 cm. Leaves 1—2 mm wide, *grooved*, slightly *curved*, hairless or sparsely hairy. Inflorescence *spike-like*, *drooping*, flowers chestnut-brown, lower bract about *as long as* inflorescence; flowering June—July. Fruit △f dark brown, globular, about *as long as* tepals. Rocks, stony and sparsely grassy places, mainly on *mountains*. B, S. Map 397.

3-d

3a-e

4-f

△ ×4

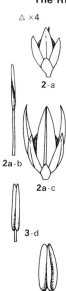

2-a

2a-b

2a-c

3-d

3a-e

1. **Great Wood-rush** *Luzula sylvatica*. Much the largest wood-rush, a stout tufted plant with numerous runners; stems medium to *tall*, to 80 cm. Leaves *5–20 mm* wide, *glossy*. Inflorescence a widely spreading forking cluster with many small heads of 2–5 brown flowers; flowering April–June. Fruit brown, egg-shaped. *Woods*, moors and rocks, mainly on *acid* soils. T. Map 398.

2. **White Wood-rush** *Luzula luzuloides*. Loosely tufted, with long runners; stems medium, to 60 cm. Leaves 3–4 mm wide. Inflorescence a loose forking cluster of small heads of *off-white* flowers, sometimes tinged red or brown, with anthers *much longer* than filaments; flowering May–June. Fruit △a globular, about *as long as* tepals. Woods, scrub and grassland, sometimes planted for ornament and more or less naturalised. T, but (B, S). Map 399. **2a Snowy Wood-rush** *L. nivea* □ has tighter clusters of snow-white flowers, not tinged red or brown, anthers no longer than filaments △b and fruit △c shorter than tepals. Woods and scrub only. F (Vosges).

3. **Alpine Wood-rush** *Luzula alpinopilosa*. Tufted; stems short to medium, to 40 cm. Leaves *c.* 3 mm wide. Inflorescence a loose forking cluster, erect or nodding, of heads of *2–5* dark brown flowers; anthers much *longer* than filaments △d; flowering June–August. Fruit egg-shaped. In wet stony places high in *mountains*, often near snow patches. F (Vosges). Map 400. **3a** *L. desvauxii* is taller, to 60 cm, and loosely tufted, with short runners and heads of 1–3 flowers. Stamen △e. G (rare). Map 401.

4. **Tundra Wood-rush** *Luzula wahlenbergii*. Tufted; stems short, to 30 cm. Leaves 3–5 mm wide. Inflorescence a forking cluster with up to 30 flowers, anthers as long as filaments, the lower bract *short* and the upper ones dark brown, fringed with *long* hairs; flowering June–August. Fruit egg-shaped. Tundra and wet grassy places. S. Map 402. **4a** *L. parviflora* □ is more loosely tufted, with broader leaves, more flowers, the lower bract much longer and the upper ones not fringed with hairs; also at wood margins. S. Map 403.

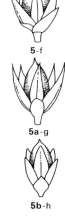

5-f

5a-g

5b-h

5. **Hairy Wood-rush** *Luzula pilosa*. Tufted, with *short* runners; stems short, to 30 cm. Leaves 5–10 mm wide. Inflorescence a forking cluster with *single flowers* on very unequal stalks, *turning down* in fruit; flowers dark brown, anthers longer than filaments; flowering *March*–May. Fruit △f yellow-green, 3-sided, *abruptly* narrowing to a short beak, *as long as* tepals. Woods and shady banks. T, except Iceland; scarce in west Ireland. T. Map 404. **5a* Southern Wood-rush** *L. forsteri* □ has narrower leaves, the stalks of the inflorescence always *erecter* and *remaining so* in fruit, anthers much longer than filaments and fruit △g narrowing more gradually to a longer point. B, F, G. Map 405. **5b** *L. luzulina* is more loosely tufted, with long runners, narrower leaves, anthers as long as filaments and fruit △h yellow-brown and longer than tepals; coniferous woods. F (Vosges). Map 406.

The Clubmoss Families Lycopodiaceae; Selaginellaceae (p. 186)

Clubmosses (pp. 184–186) are low, often prostrate, *evergreen* perennials, at first glance looking like mosses, but stouter and, unlike all mosses, with *true roots*. Their small pointed leaves have either no veins or just a midrib and overlap along the stems, but, again unlike the mosses, they have true vascular tissue in their stems and so are much *stiffer* (this and their generally larger size are among the best ways of distinguishing them from mosses). Spore-cases usually in erect cigar-shaped cones, at the base of yellow scales.

1.* Fir Clubmoss *Huperzia selago.*   The stoutest clubmoss, resembling a diminutive bushy conifer. Stems low, to 10 cm, rather stiffly *erect, forking* with all stems roughly equal; leaves untoothed, rather *sharp.* Spore-cases *on the stems,* usually near the top, at the base of the leaves, in zones alternating with sporeless zones; spores ripe July–September. *Dry* grassland on heaths, moors and mountains. T. Map 407.

2.* Marsh Clubmoss *Lycopodiella inundata.*   A slender, bright green, rather moss-like clubmoss. Stems shortly *creeping,* to 20 cm, little-branched, only *half-evergreen;* leaves *untoothed;* Cones solitary, terminal, *unstalked,* with leaf-like toothed scales; ripe August–September. Moors, wet heaths, dune slacks, especially on acid soils. T. Map 408.

3.* Interrupted Clubmoss *Lycopodium annotinum.*   Intermediate between Fir and Marsh Clubmosses (1 and 2); stems creeping like Marsh Clubmoss, but longer, to 50 cm, the branches erect, like Fir Clubmoss, but distinctively *constricted* at intervals; leaves *spreading,* minutely *toothed.* Cones as in Marsh Clubmoss, *unstalked;* ripe August–September. Moors, heaths, mountains, on *drier acid* soils. T. Map 409. **3a** *L. dubium,* with narrower stems and shorter leaves, occurs in the Arctic and Subarctic; the map does not distinguish *L. annotinum* and *L. dubium.*

4.* Stagshorn Clubmoss *Lycopodium clavatum.*   A far-creeping clubmoss, its long branched stems *trailing up to 100 cm* or more; leaves bright green, incurved, minutely toothed, ending in a *long whitish hair.* Cones *long-stalked,* or in the far north almost unstalked; ripe July–August. Moors, upland and lowland heaths, mountains. T. Map 410.

5.   Flat-stemmed Clubmoss *Diphasiastrum complanatum.*   Has far-creeping subterranean stems, *to 100 cm long,* with *flattened* branches above ground and *scale-like* leaves, some broader than others. Cones stalked; ripe August–September. Heaths, open woods. F, G, S. Map 411. **5a** *D.c.* ssp. *chamaecyparissus* has stems scarcely flattened and leaves all the same size. Map 412.

6.* Alpine Clubmoss *Diphasiastrum alpinum.*   Prostrate, with *slightly* flattened stems creeping *above ground,* to 50 cm, the branches forking to give a *flat-topped appearance;* leaves *closely appressed,* stalked on barren branches. Cones *unstalked,* greyer than leaves; ripe August–September. Upland heaths, moors, mountains. T. Map 413. **6a* Hybrid Alpine Clubmoss** *D. × issleri,* the hybrid between Flat-stemmed and Alpine Clubmosses (5 and 6), has always flattened stems, the lower leaves of barren branches narrower and unstalked, and the cone-scales broader but more sharply pointed. B (rare), F, G. Map 414.

Clubmoss Families Selaginellaceae

1: **Lesser Clubmoss** *Selaginella selaginoides.* A small moss-like plant with stems to 16 cm, slenderer and softer to the touch than Marsh Clubmoss (p. 184), but with leaves arranged irregularly, *finely toothed* and with a tiny *ligule,* which distinguishes it from true mosses. Cones less well defined, *unstalked,* yellowish; ripe July–September. Damp grassy places, dune slacks, on *basic* soils. B, G, S. Map 415.

2. **Swiss Clubmoss** *Selaginella helvetica.* Differs from Lesser Clubmoss (1) in having *flattened* stems to 20 cm, leaves of different sizes and less clearly toothed, and *stalked* cones; ripe June–July. Damp grassland, rocks. G (rare). Map 416.

3: **Mossy Clubmoss** *Selaginella kraussiana.* Moss-like carpeting plant, with flattened stems, *trailing* to 30 cm or more, and leaves arranged in 4 rows, 2 appressed and 2 longer and spreading. Cones unstalked. An escape from *greenhouse* cultivation, originating from southern Africa and maintaining itself in mild climates. (B, F, ?G). Map 417.

Quillwort Family Isoetaceae

Perennials with small tufts of quill-like leaves, each with a tiny ligule above the spore-case at its base. These spore-cases distinguish the aquatic species from barren specimens of aquatic flowering plants with similar tufts of narrow leaves, notably Pipewort *Eriocaulon septangulare,* Water Lobelia *Lobelia dortmanna,* Shoreweed *Littorella uniflora* and Awlwort *Subularia aquatica.*

4: **Quillwort** *Isoetes lacustris.* Submerged aquatic, with a 2-lobed stem and *stiff, dark green,* sometimes recurved, grass-like pointed leaves to 20 cm, of uniform width till near tip; ligule often 3-lobed. *Spineless* female spore-cases on outer leaves, much smaller male ones on inner; ripe May–July. Clear lakes, tarns and pools on *acid* soils, mainly in hill districts; often detectable by floating and grounded leaf-fragments along the shore. T. Map 418. **4a: Spring Quillwort** *I. echinospora* □ has shorter, more flaccid, *pale grass green* leaves, tapering gradually to tip; and spiny female spores. Sometimes in lowlands. Map 419. **4b** *I. tenuissima* is like *I. echinospora,* but has a 3-lobed stem and its persistent leaf-bases give it a bulbous appearance; ligule triangular. In and on the edge of shallow lakes. F (rare). Map 420.

5: **Land Quillwort** *Isoetes histrix.* *Terrestrial,* and hard to detect in the bare sandy hollows, wet only in winter, where it grows. Leaves short, wire-like, in a *2–5 cm* tuft, sometimes *recurved* right back to the ground, appearing in October and usually withered away by May. Spore-cases grey, ridged; ripe April. B (rare), F. Map 421.

Horsetail Family Equisetaceae

Perennials with hollow, ridged, roughish, leafless, jointed stems, the joints covered by toothed sheaths and often bearing whorls of ribbed, jointed, linear, leafless branches. Spores in cones at tip of some stems. Growing in more or less damp places. Note: The aquatic flowering plant Marestail *Hippuris vulgaris* (*WFBNE*, p. 292) is often miscalled horsetail. The horsetails on this page appear in spring, die back in late autumn and bear their bluntly pointed cones on stouter, pale brown stems, which appear with or before the barren green stems.

1. **Wood Horsetail** *Equisetum sylvaticum*.   The most graceful and distinctive horsetail, recalling a diminutive pale green Christmas tree. Sterile stems short to medium, to 50 cm, the numerous whorled branches being in turn finely branched and drooping at the tip; ridges 10–18, obscure. Sheaths loose, green, with 3–6 *rufous brown* teeth. Fertile stems paler, unbranched at first; ripe May, with the sterile stems. Forms colonies in woods and on moors and banks. T. Map 422.

2. **Shade Horsetail** *Equisetum pratense*.   In many ways intermediate between Wood and Field Horsetails (1 and 3). Sterile stems *short*, to 30 cm, rather stiff, slender, with numerous whorls of *thin* branches, *not branched* again, but sometimes *drooping* at the tip; ridges 8–20. Sheaths green, the teeth slender, brown with a *darker midrib*. Fertile stems unbranched at first; paler cones irregularly produced in Britain; ripe May, with the sterile stems. Forms patches in damp grassy places, especially on river banks in sandy or clayey hill districts. T. Map 423.

3. **Field Horsetail** *Equisetum arvense*.   Generally the commonest and coarsest-looking horsetail, patch-forming and much the most frequent in drier grassy and waste places; sometimes a tiresome weed. Sterile stems short to tall, to 80 cm, sometimes *sprawling, green,* usually with numerous whorls of relatively *thick* unbranched branches; ridges 6–19 on stems, 3–4 on branches. Sheaths green with *spreading* green teeth. Fertile stems whitish or *pinkish*, unbranched, soon withering; ripe April–May, before the sterile stems. Hybridises with Water Horsetail (p. 190). T. Map 424.

4. **Great Horsetail** *Equisetum telmateia*.   The largest horsetail, the *very tall* whitish sterile stems occasionally reaching 200 cm, the branches unbranched; ridges 20–40. Sheaths green, tight, with pale-edged brown teeth. Fertile stems *whitish*, unbranched, soon withering; ripe April, before the sterile stems. Forms colonies in damp places on clay soils, often near a calcareous spring. T. Map 425.

Horsetail Family Equisetaceae

The horsetails on this page bear their cones on green stems similar to the barren ones. The 3 hybrids may occur in the absence of both parents.

1.* Rough Horsetail or Dutch Rush *Equisetum hyemale.* Stems tall, to 100 cm, *stiff*, dark green, *unbranched, inflated* between the joints, *wintergreen*; ridges 10–30. Sheaths black with a green or dirty white central band, *untoothed.* Cones pointed; ripe January–April. Forms colonies in damp shady places, usually on clayey river banks. T. Map 426. **1a.* Moore's Horsetail** *E. × moorei*, the hybrid between Rough Horsetail and Boston Horsetail (2), has smoother, slenderer, uninflated yellow-green stems, only partly wintergreen, and short-lived teeth; cones ripe July–August. B.

2.* Boston Horsetail *Equisetum ramosissimum.* Stems short to tall, to 100 cm, sometimes wintergreen, with numerous *whorled branches*; ridges 8–20. Sheaths green, turning brown with a black band at base, the teeth *blackish*, pale-edged and very finely pointed. Cones pointed; ripe August–September. Damp grassy places. B (rare). F, G. Map 427.

3.* Variegated Horsetail *Equisetum variegatum.* Stems erect, to 80 cm but often much shorter, or more or less *prostrate*, often *curly*, only *branched at base*, hollow, dark green, *wintergreen*; ridges 4–10. Sheaths black above, green below, sometimes tinged *orange*, the teeth *whitish* with a black midrib. Cones pointed; ripe March–June. Damp places, often in dune slacks, on mountains and on calcareous soils. T. Map 428. **3a.* Mackay's Horsetail** *E. × trachyodon*, the hybrid of Variegated Horsetail with Rough Horsetail (1), is stouter, with longer sheaths and long black pale-edged teeth; March–April. B. **3b** *E. scirpoides* is shorter and has solid stems with only 3–4 ridges. S. Map 429.

4.* Water Horsetail *Equisetum fluviatile.* The most aquatic horsetail, often growing right *in the water*, fringing ponds, tarns or lakes, but also in swamps and carr. Stems medium to tall, to 150 cm, *dying down* in winter, usually little or *unbranched*, the central hollow occupying most of the stem, smooth, yellow-green, often tinged orange; ridges obscure, 10–30. Sheaths *tight*, green, the teeth short, black, pale-edged. Cones *blunt*; ripe June. T. Map 430. **4a.* Shore Horsetail** *E. × litorale*, the hybrid of Water Horsetail with Field Horsetail (p. 188), is more branched and less hollow, with more conspicuous ridges and looser teeth; in more marshy, less aquatic places.

5.* Marsh Horsetail *Equisetum palustre.* Stems short to medium, to 50 cm, with whorls of *branches* (except in exposed places), the central hollow *very small*; ridges 6–10. Sheaths *loose*, the teeth black with *broad white edges*, those of the branches green with black tips. Cones *pointed*; ripe June. Marshes, damp grassland. T. Map 431.

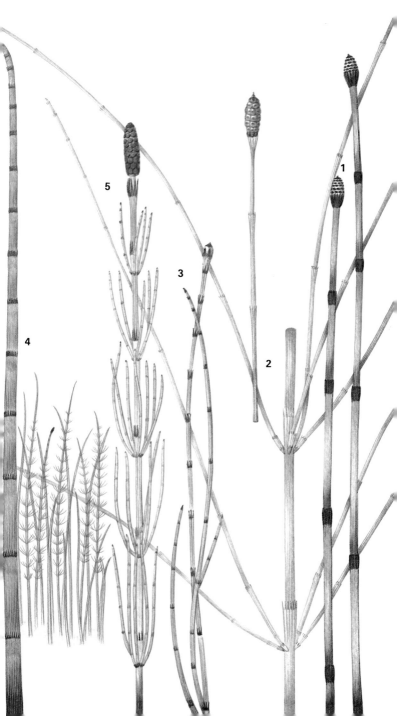

The Fern Families Filicopsida

The great majority of European ferns are perennials and bear their spores on the back of leaves (fronds) which are at least pinnately, and often multipinnately, divided, and have hairs or scales on their stems. The exceptions are Jersey Fern *Anogramma leptophylla* (p. 196), which is annual; the Adderstongues *Ophioglossum* (p. 194) and Moonworts *Botrychium* (p. 194), whose spores are in a leafless spike mimicking the flower spikes of flowering plants; Royal Fern *Osmunda* (p. 192), and Sensitive Fern *Onoclea* (p. 204), in which the terminal leaflets of some leaves become spore-cases; the Pillworts *Pilularia* (p. 216) and their allies, and the water ferns *Azolla* and *Salvinia* (p. 216), whose globular spore-cases are at the base of their leaves. Undivided leaves are found in the Adderstongues, Pillwort and Hartstongue *Phyllitis* (p. 198). (See also p. 9.)

△ ×2

Royal Fern Family Osmundaceae

1. **Royal Fern** *Osmunda regalis*. One of the *tallest* European ferns, its fine tuft of 2-pinnate leaves with *oblong* pinnae sometimes reaching 300 cm; no scales on the stalks. No green leaves, December—April. *Fertile central leaves* so thickly covered with spores, pale green at first, soon turning *golden-brown*, as to resemble the flower spike of a flowering plant; spores ripe June—August. Damp and wet, often shady places, usually on *acid* soils. T. Map 432.

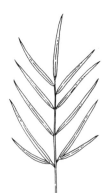

2-a

Bracken Family Hypolepidaceae

2. **Bracken** *Pteridium aquilinum*. A cosmopolitan fern that is one of the commonest in Europe and the only one that grows in *extensive*, often closed communities, often covering whole hillsides, on drier heaths, moors and in open woods. Stems up to *400 cm tall*; rootstock creeping. Leaves 3-pinnate, appearing (at first as 'shepherd's crook') from mid-April to October and persisting in a *copper-brown* dead state through the winter. Spore-cases △a covered by inrolled leaf-margins; spores ripe August—October. T. Map 433.

Pteris Family Pteridaceae

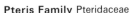

3. **Cretan Fern** *Pteris cretica* △b only. Leaves pinnate △b, up to 40 cm, with 4—7 pairs of *linear* pinnae, the lowest sometimes *forked*; stalks pale brown with dark brown scales. Spore-cases along the inrolled leaf-margins; spores ripe April—May. Walls. (B, G). **3a.** **Mediterranean Fern** *Pteris vittata* is larger, with 10 or more pairs of pinnae, the lower never forked. (B).

3-b

1

2

Adderstongue Family Ophioglossaceae

Loosely colonial. Spores borne in a separate spike or spikes arising from the base of the leaf. Adderstongues *Ophioglossum* (1–2) have undivided leaves.

△ ×½

1a-a

3a-b

3a-c

3a-d

5a-e

6a-f

△ ×$\frac{1}{10}$

1*. **Adderstongue** *Ophioglossum vulgatum.* Stems short, exceptionally to 30 cm, *solitary*, with a *single pointed oval* leaf and a single green fertile spike, both somewhat like Greater Plantain (*WFBNE*, p. 220); spores ripe June–August. Barren shoots also frequent. Plant not visible mid-November to April. Damp grassy places, often in old meadows, and dune slacks, and on calcareous soils. T. Map 434. **1a*** **Small Adderstongue** *O. azoricum* △a has stems much shorter, 3–8 cm, 2–3 together; leaves narrower, more spreading; barren shoots more numerous; and spores ripe July–August. Not visible October to early May. Damp bare places, dune slacks, near the sea. B, F. Map 435.

2*. **Least Adderstongue** *Ophioglossum lusitanicum.* Stems very short, less than 1 cm, with 1–3 *strap-like* yellow-green leaves, often detectable only by going down on hands and knees; spores ripe *April*. Barren shoots numerous. Plant not visible June–September, so the only adderstongue in leaf in *winter*. Damp bare places *by the sea.* B (rare), F. Map 436.

MOONWORTS *Botrychium* differ from Adderstongues in having the leaves *pinnate* or multipinnate, the fertile spikes *branched* and the spore-cases usually *yellowish*.

3*. **Moonwort** *Botrychium lunaria.* Stems solitary, very short to short, to 30 cm. Leaves *unstalked*, oblong, *pinnate*, with 3–9 pairs of *fan-shaped* or half-moon-shaped leaflets, with *no midrib.* Spores ripe June–August. Plant not visible September–April. Drier grassy places, including old meadows, heaths, moors, mountains and dunes, in dune slacks and on rock ledges. T. Map 437. **3a Simple Moonwort** *B. simplex* is the smallest moonwort with stems to 15 cm; leaves long-stalked, only 2–10 cm, simple △b, trefoil △c or pinnate △d, with usually 2 (rarely 3–4) pairs of roundish leaflets; spores ripe May–July. Often on limy or sandy soils. F, G, S. Map 438.

4. **Northern Moonwort** *Botrychium boreale.* Stems short, to 10–30 cm; leaves almost *unstalked*, narrowly triangular, *2-pinnate*, with *triangular* leaflets showing a *midrib*. Spores ripe July. Dry grassland. S. Map 439.

5. **Lanceolate Moonwort** *Botrychium lanceolatum.* Stems very short to short, to 25 cm; leaves almost *unstalked*, narrowly triangular to oblong, *2-pinnate*, with *lanceolate* or oblong leaflets and a *marked midrib.* Spores ripe May–July. Dry grassy, often mountainous places. S. Map 440. **5a Branched Moonwort** *B. matricariifolium* has the leaflets oval or oblong △e, the midrib often obscure; often in sandy places. F, G, S. Map 441.

6. **Leathery Moonwort** *Botrychium multifidum.* Stems short, to 25 cm; leaves *stalked*, triangular, *2–3-pinnate*, and so more 'fern-like' than most moonworts, somewhat leathery, sparsely hairy; leaflets oval, *blunt*, the midrib *conspicuous.* Spores ripe July–September. Grassland. F, G, S. Map 442. **6a Rattlesnake Fern** or **Virginian Moonwort** *B. virginianum*, the largest European moonwort, has stems 20–80 cm, and even more 'fern-like' unstalked, 3–4-pinnate △f, triangular, sparsely hairy leaves, the oval leaflets pointed. When sterile resembles Mountain Bladder Fern (p. 206). Spores ripe July–August. Upland heaths, acid grassland, open woods. S. Map 443.

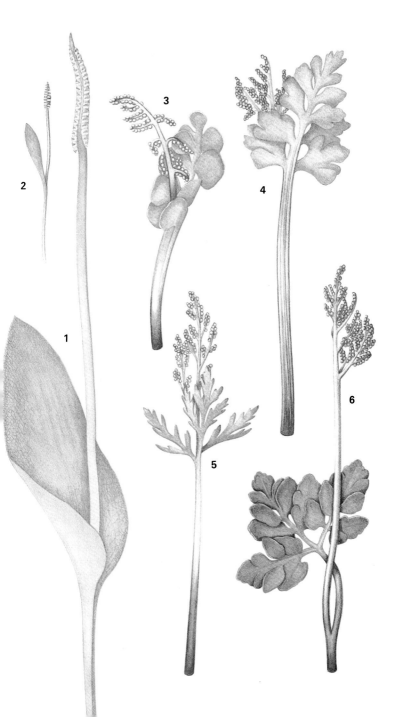

1
2
3
4
5
6

Maidenhair Fern Family Adiantaceae

△ ×2

1-a

1.* **Maidenhair Fern** *Adiantum capillus-veneris*. Leaves 2–3-pinnate, up to *c.* 15 cm, the pinnules *fan-shaped*, dying down from November to February; stalks *blackish*, with scales only at the base. Spore-cases △a at inrolled tip of pinnules; spores ripe May–August. Damp, especially *calcareous* rocks by the sea in the west, occasionally on walls and rocks inland in frost-free areas. B, F, (G). Map 444. Note: The most commonly grown greenhouse maidenhair fern, *A. cuneatum*, is taller, with larger pinnules.

Parsley Fern Family Cryptogrammaceae

2-d

2.* **Parsley Fern** *Cryptogramma crispa*. Leaves 3-pinnate, *parsley-like*, 15–30 cm, in a *bright green tuft*, dying down from December to mid April; stalks with scales only at base. Spore-cases △b on inrolled margins of narrower, linear pinnules of fertile leaves, which are erect in the middle of the tuft; spores ripe July–August. Walls, rocks, on *acid* soils, mainly in hill districts. T. Map 445.

Jersey Fern Family Gymnogrammaceae

3-c

3. **Jersey Fern** *Anogramma leptophylla*. The only *annual* true fern of our region, a delicate little fern growing on walls and banks. Leaves 2–3-pinnate, the edges flat, 3–20 cm; pinnae broadly triangular, pinnules *wedge-shaped*; dying down from July to September. Spore-cases △c *linear*, almost covering undersides of the longer inner leaves; spores ripe March–June. F. Map 446.

Filmy Fern Family Hymenophyllaceae

Small, creeping, somewhat *moss-like* plants with thin *translucent* wintergreen leaves, *forming mats* among mosses on damp rocks and tree-trunks on *acid* soils. The only European true ferns with their spore-cases in tiny pouches near the tips of the pinnae. The somewhat similar moss *Mnium undulatum* has undivided leaves without spore-cases.

4-d

4.* **Tunbridge Filmy Fern** *Hymenophyllum tunbrigense*. Leaves pinnate, 2–12 cm, *dull* green, somewhat *flattened*, the pinnae divided, with veins *not reaching* the tips. Rootstock blackish, creeping. Spore-case pouch △d minutely *toothed*, stalked; spores ripe June–July. Mainly in the lowlands. B, F. Map 447. **4a*** **Wilson's Filmy Fern** *H. wilsonii* has narrower, darker green leaves, not appearing flattened, the veins reaching the tips, and longer-stalked, untoothed pouches. Mainly in the uplands. B, F, S. Map 448.

5.* **Killarney Fern** *Trichomanes speciosum*. The only European fern with translucent 3–4-pinnate leaves, 20–35 cm long; rootstock creeping, with *blackish hairs*. Spore-cases △e in a tubular pouch, persisting as a bristle after they have gone; spores infrequent, ripe July–September. Often growing where splashed by fresh water. B, F. Map 449.

5-e

Marsh Fern Family Thelypteridaceae

△ ×1

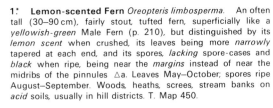

1. **Lemon-scented Fern** *Oreopteris limbosperma*. An often tall (30–90 cm), fairly stout, tufted fern, superficially like a *yellowish-green* Male Fern (p. 210), but distinguished by its *lemon scent* when crushed, its leaves being more *narrowly* tapered at each end, and its spores, *lacking* spore-cases and *black* when ripe, being near the *margins* instead of near the midribs of the pinnules △a. Leaves May–October; spores ripe August–September. Woods, heaths, screes, stream banks on *acid* soils, usually in hill districts. T. Map 450.

1-a

2. **Marsh Fern** *Thelypteris palustris* (*thelypteroides*). One of the few ferns liking really *wet* situations, in fens and marshes, and somewhat resembling a more delicate Male Fern (p. 210), but the soft, pale green 2-pinnate leaves arise *singly*, to 80–150 cm, *at intervals* from a largely underground *creeping* rootstock, their stalks almost *hairless*; leaves June–October. Spore-cases rounded, close together under the inrolled margins of the pinnules △b; spores ripe August–September. T. Map 451.

2-b

3. **Beech Fern** *Phegopteris connectilis*. A smallish fern, rarely above 30 cm high, with *triangular*, pinnate, soft, pale green leaves, the lowest pinnae being the longest, but usually *bent right down*; stalks pale whitish-green; leaves May–October. Spores without spore-cases, near the margins of the pinnules on some leaves △c, *black* when ripe; spores ripe July–September. Woods, shady rocks, avoiding both very acid and very calcareous soils; not especially, or in some districts at all, associated with beeches. T. Map 452.

3-c

Spleenwort Family Aspleniaceae

Pp. 198–202. A family of mostly rather small tufted *evergreen* ferns, bearing their spore-cases along the veins on the underside of the leaf.

4. **Hartstongue** *Phyllitis* (*Asplenium*) *scolopendrium*. The only true fern of the region that has *overwintering* tufts of undivided *strap-shaped* leaves, 10–60 cm in length and 3–6 cm in width. Spore-cases △d in diagonal rows; spores ripe August–March. Hybridises with Lanceolate Spleenwort (p. 202). Rocks, walls, hedge banks, commoner in wetter districts. T. Map 453.

4-d

Spleenwort Family Aspleniaceae

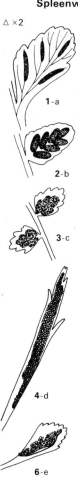

△ ×2

1.* Sea Spleenwort *Asplenium marinum.* The only fern of the region that is almost confined to *cliffs* and caves exposed to *sea spray,* only rarely, as at Killarney, south-west Ireland, growing inland. Leaves pinnate, *shiny,* rather *leathery,* bright green, evergreen, 15–30 cm, tapering at both ends with oblong pinnae, a red-brown stalk and a green midrib. Spores △a ripe mid August to mid July. B, F, S. Map 454.

1-a

2.* Maidenhair Spleenwort *Asplenium trichomanes.* Leaves evergreen, narrow, pinnate, 5–20 cm, with neat *oblong or rounded,* often slightly toothed pinnae and both stalk and midrib *blackish-brown.* Scales on the rootstock with a dark central *stripe.* Spores △b ripe August–November. Hybridises with Forked Spleenwort (4); see Hybrid Spleenwort (5). Walls and rocks. T. Map 455. **2a** *A. adulterinum,* is intermediate between Maidenhair Spleenwort and Green Spleenwort (3), with the midrib red-brown at the base, green near the tip, and some rootstock scales lacking the stripe. G, S. Map 456.

2-b

3.* Green Spleenwort *Asplenium viride.* Smaller than Maidenhair Spleenwort (2), but with leaves 5–15 cm, pinnae more distinctly *toothed,* the upper part of the stalk and all the midrib *green* and *no dark stripe* on the rootstock scales. Spores △c ripe August–November. Walls, rocks, limestone pavements, mainly in hill districts, preferring *calcareous* soils. T. Map 457.

3-c

4.* Forked Spleenwort *Asplenium septentrionale.* One of the most distinctive European ferns, with its *narrow* evergreen leaves, 4–15 cm long, hardly broader than their stalks, each usually *forked* twice, and toothed at the tip; stalks much longer than leaves, green with a red-brown base. Spore-cases △d covering almost all underside of pinnae; spores ripe August–September. Hybridises with Maidenhair Spleenwort (2); see Hybrid Spleenwort (5). Walls and rocks. T. Map 458.

4-d

5. Hybrid Spleenwort *Asplenium × alternifolium.* Intermediate between its parents, its leaves being *narrow* like Forked Spleenwort (4) but *pinnate* like Maidenhair Spleenwort (2). The pinnae are very short and narrow, the lowest being forked; stalks all dark *red-brown.* Walls and rocks.

6.* Wall-rue *Asplenium ruta-muraria.* The *smallest* spleenwort of the region, often little more than 2–5 cm high, but sometimes as much as 15 cm. Leaves evergreen, 2-pinnate, variable, the pinnae *fan-shaped,* broadest in the middle; stalks green, blackish at base. Spore-cases △e often covering whole underside of pinnae; spores ripe June–August. Walls, rocks, limestone pavements, especially on *calcareous* soils. T. Map 459.

6-e

Spleenwort Family Aspleniaceae

△ ×2

1-a

1. Rock Spleenwort *Asplenium forisiense.* Leaves 2-pinnate, pale green, wintergreen, 10–20 cm, *not tapering* towards the base, the stalk *all* red-brown, becoming green just below the leaf. Spore-cases △a near *midrib*; spores ripe July–September. Rocks, walls on *acid* soils. F. Map 460.

2. Black Forest Spleenwort *Asplenium fontanum.* The *limestone* counterpart of Rock Spleenwort (1), from which it differs in having its leaves *tapering* at each end, their stalk red-brown only *at the base,* yellowish or greenish above. Spore-cases △b. Otherwise similar. F, G. Map 461.

2-b

3.* Lanceolate Spleenwort *Asplenium billotii.* Leaves 2-pinnate, bright green, wintergreen, 15–30 cm, *not* noticeably *tapering* at the base; lowest pair of pinnae sometimes shorter and *turned down*; pinnules toothed; stalks red-brown. Spore-cases △c near *margins* of pinnules; spores ripe July–October. Rocks, walls, hedge banks, old mine-shafts, often near the sea. B, F. Map 462.

3-c

4.* Black Spleenwort *Asplenium adiantum-nigrum.* The largest spleenwort of the region, with *triangular* shiny *dark green,* wintergreen, 2–3-pinnate, leaves, 10–50 cm long; pinnules toothed, *pointed,* but blunt at the base; stalk red-brown to base of leaf. Spore-cases △d linear, over much of underside of leaves; spores ripe July–October. Rocks, walls, hedge-banks. T. Map 463. **4a* Western Black Spleenwort** *A. onopteris* has more delicate, *narrower,* yellow-green, *longer-stalked,* shinier leaves and narrower pinnules narrowly pointed at the base; stalk red-brown to well *above* base of leaf. Prefers acid rocks in Europe, but limestone in Ireland. B, F. Map 464.

4-d

5.* Rustyback *Ceterach officinarum (Asplenium ceterach).* A most distinctive small tufted fern, its dark green wintergreen *pinnately lobed* leaves, 3–25 cm long, being conspicuously *encrusted* beneath by *rusty scales* that often hide the spore-cases △e along the veins; spores ripe May–August. Walls, rocks, often on *calcareous* soils. T. Map 465.

5-e

Lady Fern Family Athyriaceae

Pp. 204–206. Spore-cases on the veins beneath the leaves.

△ ×2

1-a

1: **Lady Fern** *Athyrium filix-femina*. A large tufted fern, up to 150 cm tall, similar in general appearance to Male Fern (p. 210), but more *graceful* because its paler green 2-pinnate leaves, tapering at the base, have the pinnules *pointed* and much more deeply *toothed*; leaves April–December. Spore-cases △a *persistent, curved*; spores ripe August–November. Damper woods, rocks, hedge-banks, hillsides, streamsides, especially on *acid* soils. T. Map 466.

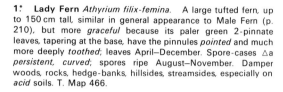

2-b

2: **Alpine Lady Fern** *Athyrium distentifolium*. A smaller version of Lady Fern (1) found on mountains, that is only certainly distinguishable by its *rounded* spore-cases △b, whose covering flap *falls off* quite early; spores ripe July–August. Usually, however, the pinnae are at an *acuter* angle, and some of the pinnules are *blunt* and often 3-lobed at the tip; leaves April–November. T. Map 467. **2a* Scottish Lady Fern** *A. flexile* □ has much shorter stalks bent right back so that the much narrower, almost parallel-sided leaves spread almost horizontally. Spores ripe May–July. B.

3. **Siberian Lady Fern** *Diplazium sibiricum*. Rootstock *creeping*, bearing at intervals *solitary*, broadly *triangular*, 25–60 cm leaves, the pinnae tapering towards the base, the lowest *bent* inwards; pinnules deeply toothed. Spore-cases small. Spruce woods. S. Map 468.

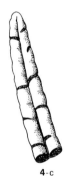

4-c

4: **Ostrich Fern** *Matteucia struthiopteris*. A *tall*, stout, tufted fern, to 150 cm, with underground runners. Barren leaves pinnate, the pinnae *pinnately lobed*, bright green, not over-wintering. Fertile leaves much shorter, in the centre of the tuft, with narrower pinnae, *not* pinnately divided, turning *brown* and so *overwintering*. Spore-cases in rows under the inrolled leaf margins △c; spores ripe June–August. Moist woods and other damp and wet places. T, but (B). Map 469.

5: **Sensitive Fern** *Onoclea sensibilis*. Leaves up to 50 cm, *solitary* along a creeping rootstock. Barren leaves *triangular*, pinnate, not overwintering, the pinnae lanceolate and coarsely toothed. Fertile leaves looking quite different, the pinnae rolled up into overwintering blackish berry-like *globules*, which contain the spore-cases and crack open to release the spores from June to October. Damp, shady places, naturalised from North America. (B, F, G).

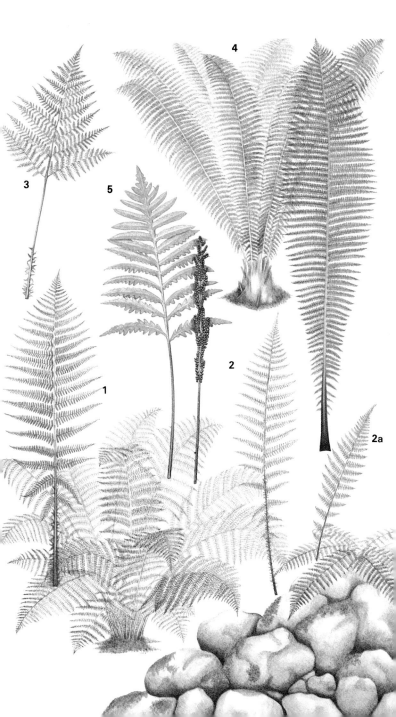

Lady Fern Family Athyriaceae

△ ×2

1-a

1. **Brittle Bladder Fern** *Cystopteris fragilis* (incl. *C. regia*). A *delicate, tufted* fern, with variable, 2–3-pinnate, 5–45 cm, *lanceolate* leaves, not much tapering at the base; stalks green but blackish at the base and with *few* scales; leaves April–November. Spore-cases △a roundish, with a *whitish* cover; spores black when ripe, June–September. Rocks, walls, mainly on *calcareous* soils. T. Map 470. **1a⁎ Dickie's Fern** *C. dickieana* is very rare (in Britain only in one or two sea caves south of Aberdeen), with shorter stalks and broader, more crowded pinnae, but only certainly distinguishable by microscopic detection of the absence of the spines on the spores that characterise Brittle Bladder Fern; spores ripe late May to mid August. B (rare), S. Map 471.

2. **Mountain Bladder Fern** *Cystopteris montana*. A delicate patch-forming fern of *mountain* rocks and woods on limestone, whose *triangular*, 3-pinnate, 10–40 cm leaves arise *singly* from its creeping rootstock; leaves May–September. Spore-cases roundish, the cover soon falling; spores ripe August–September. T. Map 472.

3. **Sudeten Bladder Fern** *Cystopteris sudetica*. Another patch-forming fern, whose leaves arise *singly* from a creeping rootstock, but the leaves are more *narrowly* triangular and *yellower*-green than Mountain Bladder Fern (2). Spore-cases roundish, the cover soon falling; spores ripe July–August. Mountain woods. G, S. Map 473.

WOODSIAS are *small tufted* ferns, rarely more than 5–6 cm high and often much smaller, with pinnate leaves whose pinnae are *pinnately lobed*. The rounded spore-cases beneath the leaves help to distinguish them from young states of other ferns. *Rocks*, on mountains except in the far north.

4-b

4. **Oblong Woodsia** *Woodsia ilvensis*. Resembles a small specimen of Brittle Bladder Fern (1), but leaves *wintergreen*, variable, 4–20 cm, not tapering to the base, the underside and midrib very *hairy* and the lower pinnae *oblong* with 4–8 pairs of lobes; stalks *brown* or red-brown, with *numerous* scales and hairs. Spore-cases △b, spores ripe July–September. B, G, S. Map 474.

5-c

5. **Alpine Woodsia** *Woodsia alpina*. Usually *smaller, paler* green and far *less hairy* than Oblong Woodsia (4), the leaf outline *broader* and coarser, the pinnae *not opposite*, with 1–4 pairs of lobes, *all triangular*. Spore-cases △c, spores ripe late June to August. Prefers basic rocks. B, S. Map 475.

6. **Smooth Woodsia** *Woodsia glabella*. Differs from both Oblong and Alpine Woodsias (4 and 5) in being almost completely *hairless* and having *narrower yellow-green* leaves, 1.5–6 cm long, the lowest pinnae *rounded*, and pale *greenish* or yellowish stalks. Spore-cases △d. S. Map 476.

6-d

Buckler Fern Family Dryopteridaceae

△ ×2

SHIELD FERNS *Polystichum* are *tufted,* up to 1 m tall, the leaves 1–3-pinnate, with sharply *pointed* pinnules ending in a *bristle.* Spore-cases roundish, on the veins beneath the leaves. The native European species may hybridise with each other.

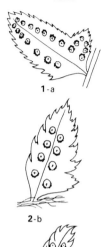

1-a

2-b

3-c

1. **Holly Fern** *Polystichum lonchitis.* Leaves evergreen, *pinnate,* 15–60 cm, narrow, stiff, *leathery,* dark green, *glossy,* the pinnae undivided, slightly curved and with spinous margins. Spore-cases △a, spores ripe August–April. Rocks, usually in hills or mountains, mainly on calcareous soils. T. Map 477. **1a*** **Sickle-leaved Holly Fern** *P. falcatum* has *broader* leaves, 30–60 cm, with longer and more markedly curved pinnae. Rocks, walls, introduced from East Asia. (B).

2. * **Hard Shield Fern** *Polystichum aculeatum.* Stout, 30–90 cm high, somewhat resembling a smaller, narrower Male Fern (p. 210). Leaves *leathery,* overwintering, 1–2-pinnate, the pinnae either pinnate or pinnately lobed and the pinnules either *unstalked* or very shortly stalked, *narrowing* down on to the midrib of the pinna, pointed, with spinous margins and a distinct terminal bristle-stalk covered with *red*-brown scales. Spore-cases △b, spores ripe July–February. Hybridises with Soft Shield Fern (4). Woods, rocks, hedge banks, often in hill districts. T. Map 478.

3. **Soft Shield Fern** *Polystichum setiferum.* Taller and *paler* green than Hard Shield Fern (3), with 30–150 cm leaves always 2-pinnate and often not overwintering. Best distinguished by its *softer* feel to the touch, and by the pinnules being blunter and less spinous, and the pinna distinctly *stalked* near its base and *not* narrowing down to the mid-rib; stalks *longer* with *golden*-brown scales. Spore-cases △c, spores ripe July–January. Hybridises with Hard Shield Fern. Woods, rocks, hedge banks, especially in wetter districts. T. Map 479.

4. **Braun's Shield Fern** *Polystichum braunii.* Intermediate in many ways between Hard Shield and Soft Shield Ferns (3 and 4). Leaves 30–90 cm, usually 2-pinnate and *soft* to the touch, but not overwintering, and pinnules un- or *very shortly* stalked and *narrowing* down on to the midrib. Differs, however, from both in its leaves being *hairy* above. Spores ripe July–August. Habitat similar, especially in hill beech and fir woods. F, G, S. Map 480.

Buckler Fern Family Dryopteridaceae

MALE FERNS and **BUCKLER FERNS** *Dryopteris* (pp. 210–212). Mostly medium to tall tufted ferns, with 2–4-pinnate leaves, the pinnules not ending in a bristle. Spore-cases kidney-shaped or roundish, on the veins beneath.

△ ×2

1-a

1. **Male Fern** *Dryopteris filix-mas*. One of the commonest larger ferns, usually with only a *single* crown on each rootstock. Leaves *half-evergreen*, 30–130 cm, tapering at both ends, 2-pinnate, the pinnules usually *pinnately lobed,* the lobes more or less triangular, rather *blunt* and with all margins more or less toothed; stalks with *pale brown* scales. Spore-cases △a usually 5–6 on each pinnule; spores ripe August–November. Hybridises with Scaly Male and Mountain Male Ferns (2 and 3). Woods, hedge-banks, rocks, screes. T. Map 481.

2. **Scaly Male Fern** *Dryopteris affinis* (*D. borreri, D. pseudomas*). Differs from Male Fern (1) in having leaves *not overwintering* (May–November), often *yellowish*-green, shinier, with a *blackish spot* at the junction of each pinna with its mid-rib, and the pinnules more or less *parallel-sided* and toothed only at the tip; stalks *covered* with *golden-brown* scales. Spores ripe August–September. Hybridises with Male Fern, the hybrid sometimes occurring in the absence of Scaly Male Fern; also with Northern Buckler Fern (p. 212). More often on acid soils. T. Map 482.

3-b

3. **Mountain Male Fern** *Dryopteris oreades* (*D. abbreviata*). Much *smaller* than typical Male Fern (1), rarely more than 50 cm, with *several* crowns on each rootstock, the leaves appearing stiff, rather *crinkly* and slightly fragrant when crushed; scales *greyish*; leaves May–November. Spore-cases △b, only 1–2 under each pinnule; spores ripe August–October. Hybridises with Male Fern. *Mountain* slopes and screes. T. Map 483.

4-c

4. **Rigid Buckler Fern** *Dryopteris submontana* (*D. villarii*). Leaves stiff, 20–60 cm, *greyish mealy*, 2–3-pinnate, sometimes almost 3-pinnate, triangular, usually with many yellowish hairs, *balsam-scented*, the pinnules with pointed teeth; leaves May–November; stalks yellowish with a blackish base and pale brown scales. Spore-cases △c, spores ripe July–September. Rocks, *limestone pavements*, usually on calcareous soils. B, F, G. Map 484.

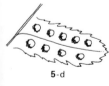

5-d

5. **Crested Buckler Fern** *Dryopteris cristata*. Creeping, with small tufts. Leaves 30–90 cm, 1–2-pinnate, *narrow* and more or less parallel-sided, pale green, the pinnae *short* with sharply pointed teeth; stalks with pale brown scales; leaves June–December. Outer sterile leaves half-erect, twice as long as stalk; inner fertile leaves erect, only slightly longer than stalk. Spore-cases △d, spores ripe August–September. Fens, bogs, *wet places*. T. Map 485.

Buckler Fern Family Dryopteridaceae

△ ×2

1-a

1. **Broad Buckler Fern** *Dryopteris dilatata* (*D. austriaca*). One of the commoner European ferns, with leaves isosceles-triangular, rather spreading, 10–150 cm, *dark green, 3-pinnate*; stalks with scales all *dark* or *dark-centred*; leaves April–November. Hybridises with Northern and Narrow Buckler Ferns (2 and 3). Spore-cases △a, spores ripe July–August. Woods, hedge banks, heaths, mountain slopes, preferring *acid* and dryish soils. T. Map 486.

2. **Northern Buckler Fern** *Dryopteris expansa* (*D. assimilis*). Differs from Broad Buckler Fern (1) in having paler, *yellower*-green, more *broadly* triangular leaves, with the lowest pinnule of the lowest pinna at least *half as long* as the pinna (in Broad Buckler Fern it is usually less than half as long); scales *rufous* brown; leaves May–November. Hybridises with Broad Buckler Fern. Spores ripe August–September. Mountains, wet woods. B, G, S. Map 487.

3-b

3. **Narrow Buckler Fern** *Dryopteris carthusiana*. Differs from Broad Buckler Fern (1) in having *narrower*, more parallel-sided, usually shorter and erecter, pale to *yellowish*-green, 2-pinnate leaves and paler brown scales with *no dark centre*; leaves June–October. Spore-cases △b, spores ripe August–September. Hybridises with Broad Buckler Fern. Damp woods, bogs and rather *wet* places. T. Map 488.

4. **Hay-scented Buckler Fern** *Dryopteris aemula*. Named from the fragrance of the *freshly dried* leaf. Leaves more broadly triangular than Broad Buckler Fern (1), 15–60 cm, *bright* green, *wintergreen*, 3-pinnate, the lowest pair of pinnae often noticeably larger than the others, the pinnules *crisped*; scales on stalks uniformly *red*-brown. Spores ripe August–October. Woods, hedge banks, shady rocks. B, F. Map 489.

5. **Fragrant Buckler Fern** *Dryopteris fragrans*. Leaves *short*, 10–20 cm, narrowly triangular, pinnate, the pinnae *pinnately lobed* and *fragrant* from numerous tiny glands; stalks with uniformly brown or *red-brown* scales. Rocks, screes; Arctic Finland.

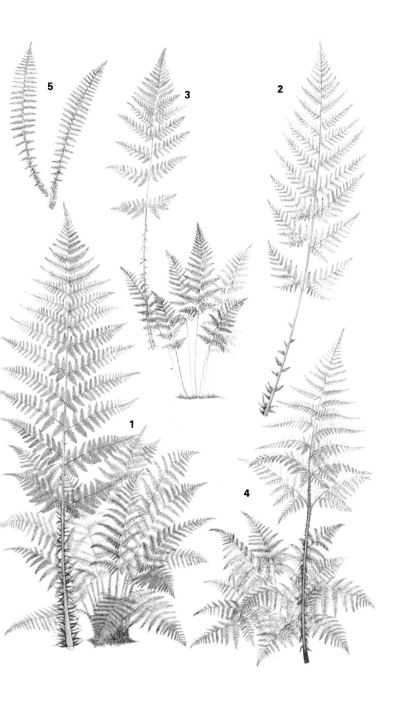

Buckler Fern Family Dryopteridaceae

1-a

1: **Oak Fern** *Gymnocarpium dryopteris.* Leaves growing *singly* from a creeping rootstock, 10–40 cm, *bright* green, *hairless,* 3-pinnate, the lowest pair of pinnae often as long as the rest of the leaf, making the whole *broadly triangular;* held *horizontally* on blackish stalks with few, pale brown scales; leaves May–September. The leaves in bud are rolled into *three* balls. Spore-cases △a roundish, on veins near the margins; spores ripe July–August. Damp woods, shady rocks, mountain screes, avoiding calcareous soils. T. Map 490.

2: **Limestone Fern** *Gymnocarpium robertianum.* The limestone counterpart of Oak Fern (1), from which it differs in its often *taller,* 15–60 cm, more *upright,* larger, *duller* green leaves which, with their stalks, are slightly *hairy* and (unlike Oak Fern) *fragrant* when bruised; leaves in bud rolled into a *single* ball. Less shade-loving, on *limestone* rocks and screes. T. Map 491.

Hard Fern Family Blechnaceae

3: **Hard Fern** *Blechnum spicant.* A distinctive tufted fern, the lanceolate, *pinnate,* overwintering leaves 10–70 cm, with fairly uniform bluntish oblong pinnae joined at the base. Outer leaves barren and more or less spreading, inner fertile ones erect and with narrower pinnae, the spore-cases linear beneath them; spores ripe August–November. Woods, heaths, moors, *avoiding* limy soils. T. Map 492.

Polypody Family Polypodiaceae

POLYPODIES *Polypodium.* Leaves *overwintering,* arising from a creeping rootstock, up to 50 cm, usually much less, *pinnate* with oblong pinnae, more or less joined at the base. Spore-cases rufous, in short rows on the veins beneath. The 3 species frequently hybridise. Woods, often on mossy tree trunks, rocks, walls, often in drier districts than other ferns.

4-b

4: **Common Polypody** *Polypodium vulgare.* One of the commoner ferns of the region, the lanceolate, rather leathery leaves appearing in May, their pinnae fairly *blunt* and mostly more or less *equal* in length, the lowest pair usually *not* projecting forward, so that the base of the leaf appears more or less *rectangular.* Spore-cases △b *roundish;* spores ripe August–March. Preferring *acid* soils. T. Map 493.

5-c

5: **Western Polypody** *Polypodium interjectum.* Intermediate in many respects between Common and Southern Polypodies (4 and 6), from which it is derived, resembling Common Polypody in having rather leathery leaves and producing new ones early (June), and Southern Polypody in having its pinnae *pointed,* the lowest usually projecting *forwards,* and the spore-cases △c roundish to *elliptical.* Leaf-shape is intermediate and distinctive, the longest pinnae being somewhat below the middle, so that the leaf *tapers* at *both* ends. Spores ripe September–February. Prefers *basic* soils. B, F, G. Map 494.

6-d

6: **Southern Polypody** *Polypodium australe:* see p. 216.

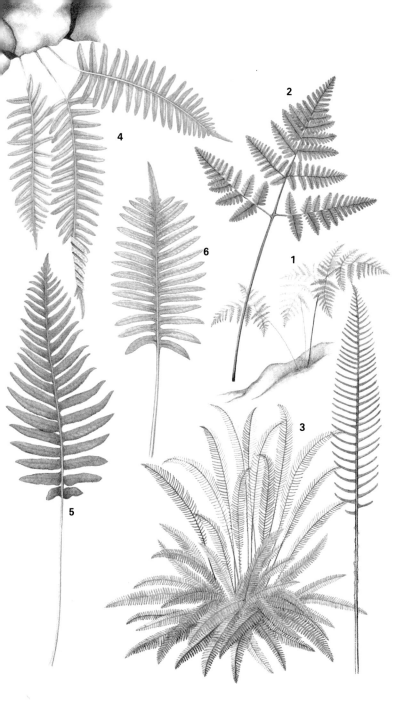

Clover Fern Family Marsileaceae

Most unfern-like little *semi-aquatic* ferns, with leaves arising in small tufts from a creeping rootstock, along which lie the spore-cases, looking like small *blackish pills* or nodules. Map 495.

1. Clover Fern *Marsilea quadrifolia.* Leaves like *4-leaved clovers*, on 7–20 cm stalks. Spores ripe September–October. Muddy ponds, marshes, ditches. F, G. Map 496.

2. **Pillwort** *Pilularia globulifera.* Leaves *thread-like*, mat-forming, yellow-green, 5–10 cm, showing their fern affinities when young by being *coiled.* In or by shallow muddy fresh water, marshes, wet heaths, often on *acid* soils. Spores ripe June–September. Increasingly rare in Europe. T. Map497.

Floating Fern Family Salviniaceae

3. Floating Fern *Salvinia natans.* More like a large duckweed than a fern, being entirely *floating* with leaves in *whorls of 3*, 2 of them floating, undivided and hairy, the third submerged, root-like and pinnate with very fine pinnae, almost *feather-like.* Spore-cases at the base of the submerged leaves. Ponds, ditches, marshes. G. Map 498.

Water Fern Family Azollaceae

Another family of small floating ferns, but less duckweed-like and slightly more fern-like because of their branched stems. Spore-cases under the first pinna on each branch. Roots thread-like. Ponds and other still fresh water, often covering the surface like and sometimes mixed with duckweeds (see *WFBNE*, p. 292).

4. **Water Fern** *Azolla filiculoides.* *Elliptical* in shape, up to 10 cm long, with the oblong leaves *overlapping* along the stems, blue-green, soon *reddening,* and a blaze of red in autumn; the upper lobe blunt with a *broad* whitish margin. Spores ripe June–September. Introduced from tropical America. (B, F, G). Map 499.

5. Lesser Water Fern *Azolla caroliniana.* *Roundish* in shape, up to 1 cm across, with pale green roundish leaves, *not* reddening and *scarcely* overlapping, the upper lobe almost pointed, with a *narrow* whitish margin. Introduced from North America. (F, G). Map 500.

---

6. **Southern Polypody** *Polypodium australe* (see p. 215). Differs from Common Polypody (4) in having leaves shorter, 5–15 cm, broader, *yellower*-green and not leathery, the new ones appearing in *August*, with *pointed* pinnae, longer towards the base and the lowest pair often held upwards, away from the mid-rib, so that the leaf appears *triangular*, and in its *oval* or elliptical spore-cases △d, spores ripe December–May. Much prefers *basic* soils. B, F. Map 495.

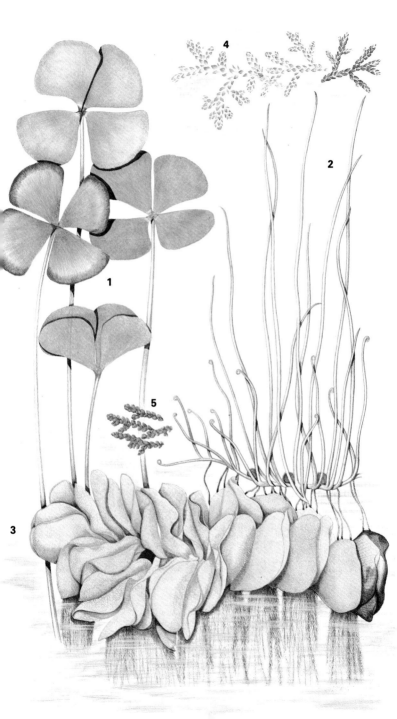

Appendix 1

Festuca ovina agg.

The taxonomy of the genus *Festuca* is extremely complex and on p. 48 we have not attempted to give full details of the species resembling *F. ovina*. These are listed below, but this is only a provisional outline of the group. Most of the characters used by taxonomists to delimit these species are obscure and frequently require examination of the internal leaf anatomy. The status of many as other than locally segregated populations is doubtful.

1. *F. costei* group comprises *F. costei*, *F. hervieri* and *F. patzkiei*. All are found in west Europe (but not in Britain so far), the last-named only in France. They have less thread-like leaves (to 0.5 mm), indistinct auricles, and in the first two the leaf-sheaths are not open to the base.

2. *F. ovina* group consists of *F. tenuifolia* (T, p. 48), *F. ovina* (T, p. 48) *F. ophioliticola* (F, rare), *F. armoricana* (F, rare), *F. heteropachys* (F, G) *F. guestfalica* (B, F, G), *F. lemanii* (B, F), *F. longifolia* (B, F), *F. polesica* (G, S), *F. pallens* (F, G), *F. huonii* (F, rare), *F. vivipara* (B, S; p. 48), and *F. indigesta* (Ireland? One old record, probably in error). *F. guestfalica* and *F. lemanii* are both recorded from Britain but as they are not reliably distinct in published descriptions, they are best regarded as a single, variable entity, characterised by their rough, less narrow leaves (over 0.6 mm wide).

3. *F. valesiaca* group, characterised by their flattened leaves (in cross-section — roll between fingers) is mainly central European. It includes *F. valesiaca* (F, G), *F. pseudovina* (G), *F. rupicola* (F, G), *F. trachyphylla* ((B), F, G, (S)) and *F. duvalii* (F, G). *F. trachyphylla* is said to be widely naturalised and on the increase: it is the only *ovina*-like fescue known to occur in Britain which has flattened leaves.

There is considerable doubt about the status of some of these species, but no doubt that intensive work on the sheep's fescues of any part of Britain would reveal other combinations of characters than those described. There is even evidence that typical *F. ovina* is, in fact, a much rarer plant than has hitherto been thought to be the case.

Appendix 2

Carex flava agg.

As we go to press, Dr B. Schmid of the University College of North Wales and the University of Zürich has published a revised nomenclature for this group (*Watsonia*, 14: 309–19). As it is too soon to know whether it will gain acceptance, we give it below, instead of changing the text.

The several species of the *Carex flava* group (see p. 150) have been amalgamated as follows:

2. *Carex flava* remains *C. flava.*
2a. *C. jemtlandica* is considered to be a mixture of *C. flava* and *C. lepidocarpa.*
3. *C. lepidocarpa* becomes *C. viridula* ssp. *brachyrhyncha.*
4. *C. demissa* becomes *C. viridula* ssp. *oedocarpa.*
5. *C. serotina* becomes *C. viridula* ssp. *viridula.*
5a. *C. bergrothii* becomes *C. viridula* ssp. *viridula* var. *bergrothii.*

Glossary

Acid soils have few basic minerals such as calcium and magnesium and are typically formed on rocks such as sandstones and granites. Peats are normally acid, since they are composed almost entirely of unrotted plant remains and have no reserve of minerals; only those flushed by base-rich water will not be acid.

Aggregate: A group of closely related species, often only distinguished with difficulty.

Annual plants live for a year or less. They are usually shallow-rooted and rarely reproduce vegetatively by rhizomes or stolons.

Anthers are the pollen sacs borne on filaments. They are the male organs of the flower and are often yellow or purple. In many grasses, sedges and rushes, they are the most conspicuous part of the flower.

Appressed: flattened against the stem or other organ.

Auricles are small flaps or horns which project from the leaf-base and may encircle the stem. A good character in some grasses (e.g. ryegrass) and sedges.

Awn: a bristle that is characteristic of the florets of many grasses. It arises from the lemma (q.v.), either as a continuation of the central nerve or from the back, and may be very long (e.g. in barleys) and sometimes hairy.

Base-rich or basic soils have large amounts of basic minerals (mainly calcium) and so are only slightly acid, neutral or even alkaline.

Beak: the tip of the fruit of sedges is often drawn out into a beak.

Blade: the flat part of a leaf.

Bogs occur on wet, more or less acid peat, and are often dominated by *Sphagnum* mosses and sedges.

Bracts are more or less leafy organs which occur just below the spikes in sedges, where they may be very long. They also occur, reduced to small scales, in the florets of both sedges and grasses.

Bulbils are tiny bulb-like propagules which occur either in the axils of leaves or in place of flowers, and can give rise to new plants.

Calcareous soils are formed over chalk and limestone, and so are extremely base-rich (q.v.). They are never more than slightly acid at the surface, and typically have a rather rich flora.

Carr is a woodland of alder or willow formed in very wet conditions.

Casual plants appear only irregularly and are not native to the area.

Cluster: a loose group of flowers or spikes.

Cones: in horsetails and clubmosses the fertile, spore-bearing region, more or less clearly differentiated from the stem.

Cultivar: a variety (q.v.) produced in cultivation by breeding.

Dichotomous: branching by regular, equal forking.

Dunes are areas of wind-blown, usually calcareous shell-sand near the sea, with areas of damp ground termed 'slacks' in between.

Female flowers contain styles only, no stamens.

Fens occur on base-rich peat, in contrast to bogs which are acid; the source of the bases is always infiltrating water from the surrounding land, since peat has no mineral reserves. Poor fens are nutrient-poor or acid, changing to bogs.

Filament: the thread on which the anther (q.v.) is borne.

Floret: generally a small flower, but used specifically to refer to the flowers of grasses and sedges, which are aggregated into spikes or spikelets (see pp. 12–16).

Flushes are areas where water seeps out of the ground, before it has resolved itself into a stream. Common in upland areas and usually moss-dominated; base-rich alpine flushes have a very exciting flora.

Frond: the 'leaf' of a fern, arising directly from the rhizome or stock, including both its stalk and the more or less divided blade.

Fruit: a seed and its surrounding parts. In sedges the fruit is a nut that is borne inside a flask called the utricle: this whole structure is often loosely referred to as the fruit.

Glaucous: bluish or greyish, usually due to waxy blooms.

Glumes are small bracts which occur at the base of the spikelets of both grasses and sedges (see pp. 12–16).

Head: a more or less tight group of flowers, borne at the end of a stem.

Heaths are dry, heather- or gorse-dominated areas in lowland regions, with rather acid soils. Wet heaths lie between bogs and heaths.

Hermaphrodite: a flower with both male and female parts.

Hybrids result from the interbreeding of two distinct species and are very often sterile.

Inflorescence: a group of flowers with their associated bracts, branches, etc.

Introduced plants are not native, but have been brought in by man, accidentally or deliberately.

Lanceolate: spear-shaped, narrowly oval and pointed.

Lax: of a flowerhead with the flowers or spikes well spaced.

Lemma: one of the scale-like bracts in a grass floret (see p. 15).

Ley: grass grown as a crop.

Linear: almost parallel-sided.

Ligule: a small flap at the junction of leaf and stem in grasses and sedges.

Male flowers contain stamens only and no styles.

Margin: edge.

Marshes are wet habitats on mineral soils, though often with a thin layer of peat.

Midrib: the central vein or nerve of a leaf or bract.

Moors are the upland counterpart of heaths (q.v.) and are typically dominated by heather, though bilberry, grasses or mosses may be important.

Node: the point of origin of a leaf on the stem; often swollen or hard. In creeping plants, adventitious roots often arise at nodes.

Nut: the fruit of sedges and rushes – a hard-coated, angled fruit. Generally a 1-seeded fruit that does not split at maturity.

Ovary: see style.

Panicles are branched inflorescences in which there is no terminal flower, because of continuous growth; the youngest flowers are therefore at the top.

Peat is a soil wholly composed of the un- or partially decomposed remains of plants which once grew on the site. It typically occurs in waterlogged conditions in which decomposition is very slow. Peat may preserve a record of thousands of years of vegetation history. Peats are found in fens and bogs (q.v.).

Perennials are plants that survive for more than a single growing season. There are no biennials (plants that germinate and grow vegetatively for one year and flower and die in the next) in this book; the category is an artificial one anyway.

Pinna: the primary division of a pinnate leaf (q.v.).

Pinnate leaves are regularly divided into pairs of leaflets. In 2-pinnate leaves, these leaflets (the pinnae) are themselves divided pinnately, these secondary divisions being called pinnules. In ferns 3- and 4-pinnate leaves occur, though the final division is often only partial (cf. pinnatifid).

Pinnatifid leaves are pinnate but the divisions do not go as far as the midrib.

Pinnule: see Pinnate.

Pith: spongy tissue that fills otherwise hollow stems; found in the stems of many rushes.

Recurved: curved backwards or downwards.

Rhizomes are underground stems, from which shoots arise, sometimes swollen with stored food. They may be horizontal and far-creeping, short, or even upright.

Rosette: a more or less flattened, rose-like group of leaves arising directly from the roots or an underground stem.

Runners are horizontal, above-ground stems, often rooting at the nodes.

Scales are small flaps of tissue, often brown and papery; not leaf-like.

Sheath: the lower part of the leaf which is wrapped around the stem, often forming a complete tube.

Shy flowerers are plants that do not flower every year.

Slack: a damp area in a sand-dune system.

Spike: a simply branched inflorescence, with unstalked or short-stalked flowers arising from a central axis. Sedges have 1 or several spikes in their inflorescences.

Spikelet: the basic unit of the inflorescences of grasses and sedges. It contains 1 (sedges *Carex* and some grasses) or several (most grasses and other Cyperaceae) florets.

Spores are small reproductive bodies, the most prominent part of the life cycle in flowerless plants such as ferns.

Stamens comprise anthers and filaments (q.v.).

Sterile stems are those that bear only leaves and no flowers.

Stigma: see Style.

Stolon: see Runners.

Style: the filament leading from the ovary of the female organ to the stigma, the receptive surface on which the pollen germinates. The pollen-tube grows down the style to fertilise the egg.

Subspecies: the division in the classification of organisms immediately below the species. Subspecies are morphologically distinct and usually geographically separated, but interbreed freely if brought together, unlike most species.

Tepals are the sepals and petals of rush flowers, so-called because the two are indistinguishable (cf. p. 12).

Trefoil: a compound leaf with 3 leaflets.

Variety: a morphologically distinct form of a plant, of lower rank than a subspecies and not geographically defined. Typically varieties occur as isolated individuals in a population, and often erratically.

Viviparous plants produce tiny plantlets in the inflorescence in place of some or all of the flowers. Such reproduction is wholly vegetative.

Waste places: areas much disturbed by man, but not cultivated.

Whorl: where more than two organs arise at the same point on a stem.

Wintergreen: of a plant that remains green through the winter.

The Maps

The Maps

The maps have been compiled on the same basis as those in *An Atlas of the Wild Flowers of Britain and Northern Europe* (see Further Reading, p. 250), and cover a wider area than that used for determining the species to be included (see Introduction, p. 5). The sources used are largely the same as for that companion volume. The only species in the text not mapped here are those which are casuals in the whole area and so have no fixed sites. Well-established introductions are not distinguished from native records here. The maps vary in reliability, being best for Britain and Scandinavia and least reliable for western France. That part of the range shown on the maps which lies outside the scope of this book (Poland, Czechoslovakia, East Germany, Austria) has been included here for the sake of completeness; available information is often very inadequate.

Green represents the core of each species's range, that part in which it is most likely to be found.

Yellow denotes the area in which the species occurs less frequently. Where information is poor this colour may represent a small number of scattered localities which cannot be easily localised.

The distinction between the two colours is relative, and cannot be compared between species.

The thick broken line indicates the southern and eastern limits of the area covered by the guide (see p. 5). Thin dotted lines indicate political boundaries.

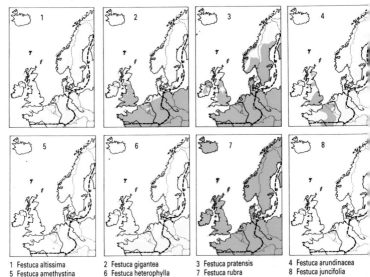

1 Festuca altissima
5 Festuca amethystina

2 Festuca gigantea
6 Festuca heterophylla

3 Festuca pratensis
7 Festuca rubra

4 Festuca arundinacea
8 Festuca juncifolia

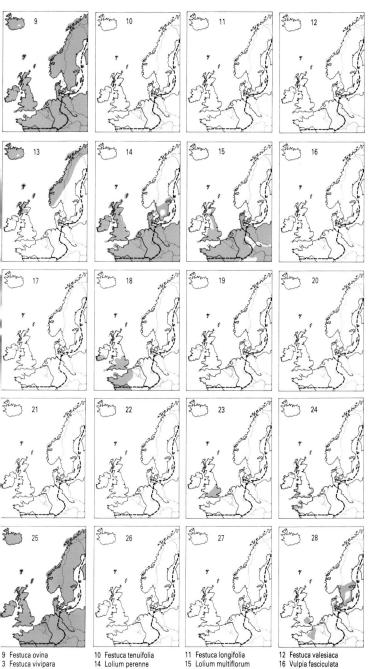

9 Festuca ovina
3 Festuca vivipara
7 Vulpia membranacea
1 Vulpia unilateralis
5 Poa annua

10 Festuca tenuifolia
14 Lolium perenne
18 Vulpia bromoides
22 Micropyrum tenellum
26 Poa supina

11 Festuca longifolia
15 Lolium multiflorum
19 Vulpia myuros
23 Desmazeria rigida
27 Poa infirma

12 Festuca valesiaca
16 Vulpia fasciculata
20 Vulpia ciliata
24 Desmazeria marina
28 Poa compressa

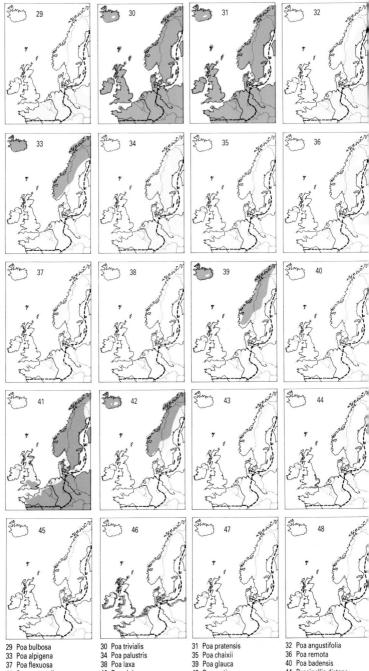

29 Poa bulbosa
33 Poa alpigena
37 Poa flexuosa
41 Poa nemoralis
45 Puccinellia fasciculata

30 Poa trivialis
34 Poa palustris
38 Poa laxa
42 Poa alpina
46 Puccinellia maritima

31 Poa pratensis
35 Poa chaixii
39 Poa glauca
43 Poa arctica
47 Puccinellia phryganodes

32 Poa angustifolia
36 Poa remota
40 Poa badensis
44 Puccinellia distans
48 Puccinellia rupestris

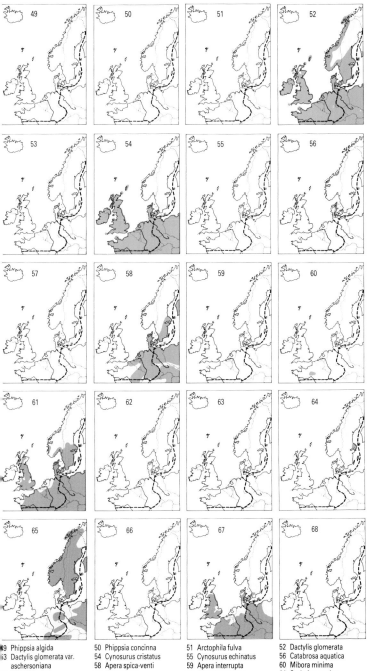

49 Phippsia algida
53 Dactylis glomerata var.
   aschersoniana
57 Cinna latifolia
61 Briza media
65 Melica nutans

50 Phippsia concinna
54 Cynosurus cristatus
58 Apera spica-venti
62 Briza minor
66 Melica picta

51 Arctophila fulva
55 Cynosurus echinatus
59 Apera interrupta
63 Sesleria albicans
67 Melica uniflora

52 Dactylis glomerata
56 Catabrosa aquatica
60 Mibora minima
64 Sesleria caerulea
68 Melica ciliata

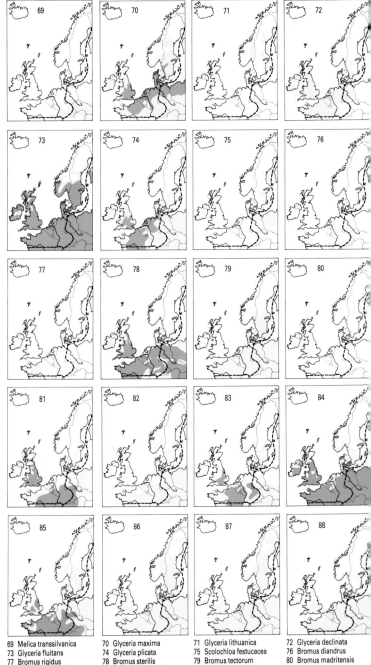

69 Melica transsilvanica
73 Glyceria fluitans
77 Bromus rigidus
81 Bromus ramosus
85 Brachypodium pinnatum

70 Glyceria maxima
74 Glyceria plicata
78 Bromus sterilis
82 Bromus benekenii
86 Bromus arvensis

71 Glyceria lithuanica
75 Scolochloa festucacea
79 Bromus tectorum
83 Bromus erectus
87 Bromus secalinus

72 Glyceria declinata
76 Bromus diandrus
80 Bromus madritensis
84 Brachypodium sylvaticum
88 Bromus commutatus

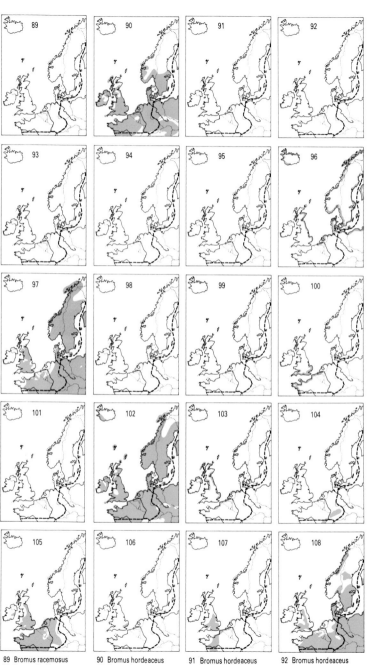

89 Bromus racemosus
93 Bromus lepidus
97 Elymus caninus
101 Elymus pungens
105 Hordeum murinum

90 Bromus hordeaceus
94 Bromus japonicus
98 Elymus mutabilis
102 Elymus repens
106 Hordeum marinum

91 Bromus hordeaceus
   thominii
95 Bromus squarrosus
99 Elymus alaskanus
103 Elymus farctus
107 Hordeum secalinum

92 Bromus hordeaceus
   ferronii
96 Elymus arenarius
100 Elymus pycnanthus
104 Hordelymus europaeus
108 Avenula pubescens

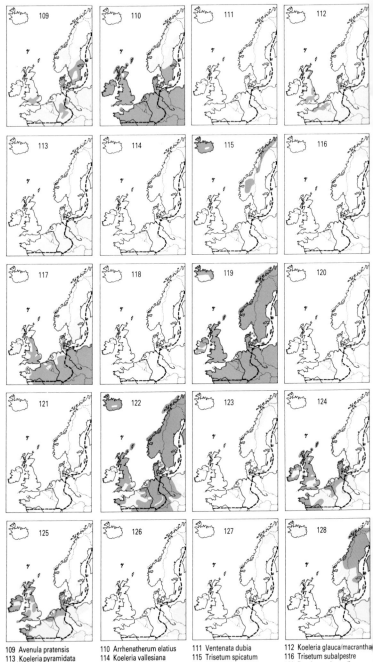

109 Avenula pratensis
113 Koeleria pyramidata
117 Trisetum flavescens
121 Deschampsia setacea
125 Aira caryophyllea

110 Arrhenatherum elatius
114 Koeleria vallesiana
118 Lagurus ovatus
122 Deschampsia flexuosa
126 Hierochloe australis

111 Ventenata dubia
115 Trisetum spicatum
119 Deschampsia cespitosa
123 Vahlodea atropurpurea
127 Hierochloe alpina

112 Koeleria glauca/macrantha
116 Trisetum subalpestre
120 Deschampsia media
124 Aira praecox
128 Hierochloe odorata

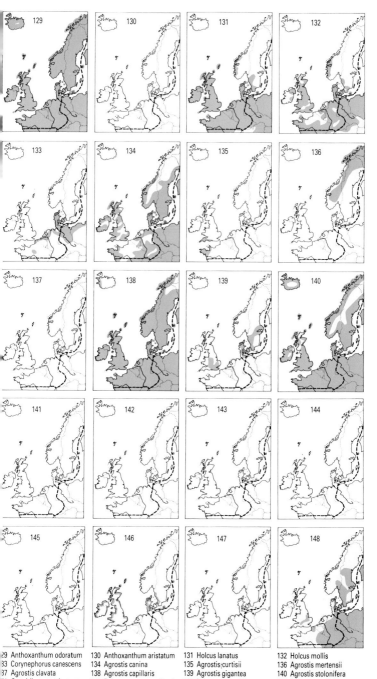

129 Anthoxanthum odoratum
133 Corynephorus canescens
137 Agrostis clavata
141 Gastridium ventricosum
145 × Agropogon littoralis

130 Anthoxanthum aristatum
134 Agrostis canina
138 Agrostis capillaris
142 Polypogon monspeliensis
146 Ammophila arenaria

131 Holcus lanatus
135 Agrostis curtisii
139 Agrostis gigantea
143 Polypogon maritimus
147 × Ammocalamagrostis
       baltica

132 Holcus mollis
136 Agrostis mertensii
140 Agrostis stolonifera
144 Polypogon viridis
148 Calamagrostis epigeios

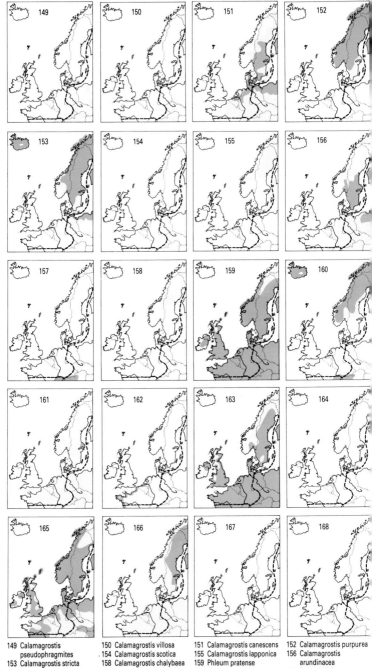

149 Calamagrostis
   pseudophragmites
153 Calamagrostis stricta
157 Calamagrostis varia
161 Phleum phleoides
165 Alopecurus geniculatus

150 Calamagrostis villosa
154 Calamagrostis scotica
158 Calamagrostis chalybaea
162 Phleum arenarium
166 Alopecurus aequalis

151 Calamagrostis canescens
155 Calamagrostis lapponica
159 Phleum pratense
163 Alopecurus pratensis
167 Alopecurus bulbosus

152 Calamagrostis purpurea
156 Calamagrostis
   arundinacea
160 Phleum alpinum
164 Alopecurus arundinaceus
168 Alopecurus alpinus

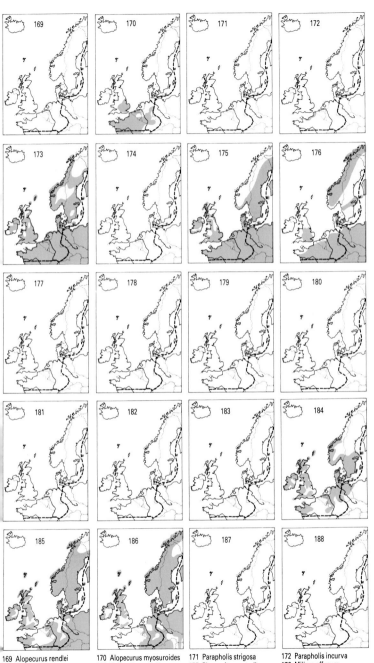

169 Alopecurus rendlei
173 Phalaris arundinacea
177 Milium vernale
181 Stipa joannis
185 Molinia caerulea

170 Alopecurus myosuroides
174 Coleanthus subtilis
178 Stipa pulcherrima
182 Stipa capillata
186 Nardus stricta

171 Parapholis strigosa
175 Phragmites australis
179 Stipa bavarica
183 Achnatherum
   calamagrostis
187 Eragrostis pilosa

172 Parapholis incurva
176 Milium effusum
180 Stipa tirsa
184 Danthonia decumbens
188 Eragrostis minor

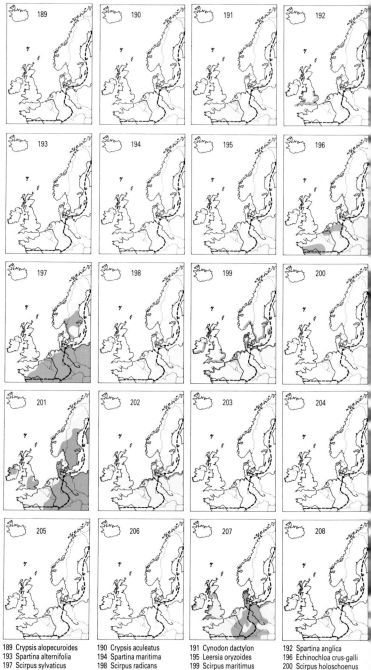

189 Crypsis alopecuroides
193 Spartina alternifolia
197 Scirpus sylvaticus
201 Scirpus lacustris
205 Scirpus triqueter

190 Crypsis aculeatus
194 Spartina maritima
198 Scirpus radicans
202 Scirpus lacustris ssp
     tabernaemontani
206 Scirpus supinus

191 Cynodon dactylon
195 Leersia oryzoides
199 Scirpus maritimus
203 Scirpus pungens
207 Scirpus setaceus

192 Spartina anglica
196 Echinochloa crus-galli
200 Scirpus holoschoenus
204 Scirpus mucronatus
208 Scirpus cernuus

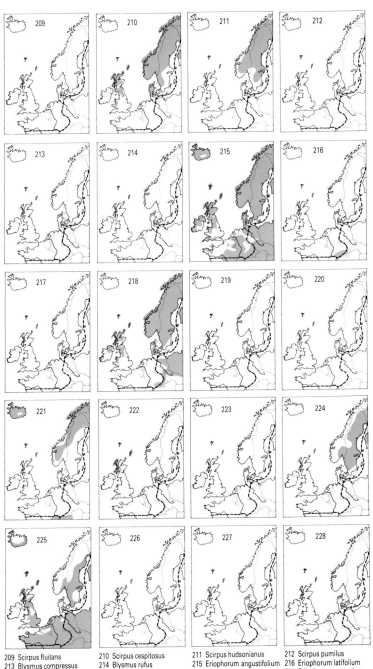

209 Scirpus fluitans
213 Blysmus compressus
217 Eriophorum gracile
221 Eriophorum scheuchzeri
225 Eleocharis palustris

210 Scirpus cespitosus
214 Blysmus rufus
218 Eriophorum vaginatum
222 Eleocharis quinqueflora
226 Eleocharis austriaca

211 Scirpus hudsonianus
215 Eriophorum angustifolium
219 Eriophorum
    brachyantherum
223 Eleocharis parvula
227 Eleocharis mamillata

212 Scirpus pumilus
216 Eriophorum latifolium
220 Eriophorum russeolum
224 Eleocharis acicularis
228 Eleocharis uniglumis

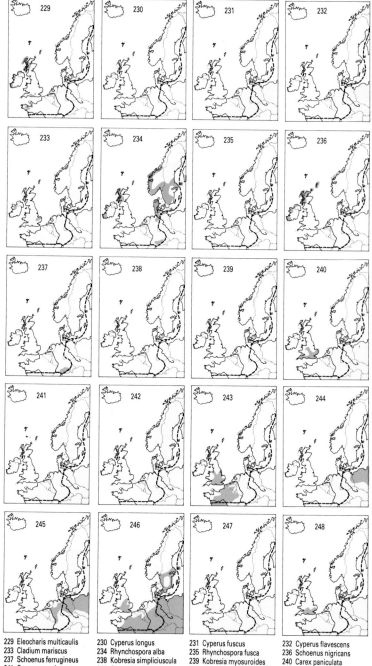

229 Eleocharis multicaulis
233 Cladium mariscus
237 Schoenus ferrugineus
241 Carex appropinquata
245 Carex elongata

230 Cyperus longus
234 Rhynchospora alba
238 Kobresia simpliciuscula
242 Carex diandra
246 Carex spicata

231 Cyperus fuscus
235 Rhynchospora fusca
239 Kobresia myosuroides
243 Carex otrubae
247 Carex muricata

232 Cyperus flavescens
236 Schoenus nigricans
240 Carex paniculata
244 Carex vulpina
248 Carex divulsa

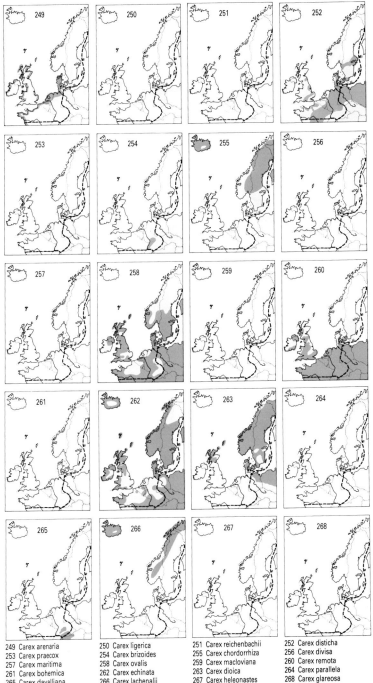

249 Carex arenaria
253 Carex praecox
257 Carex maritima
261 Carex bohemica
265 Carex davalliana
250 Carex ligerica
254 Carex brizoides
258 Carex ovalis
262 Carex echinata
266 Carex lachenalii
251 Carex reichenbachii
255 Carex chordorrhiza
259 Carex macloviana
263 Carex dioica
267 Carex heleonastes
252 Carex disticha
256 Carex divisa
260 Carex remota
264 Carex parallela
268 Carex glareosa

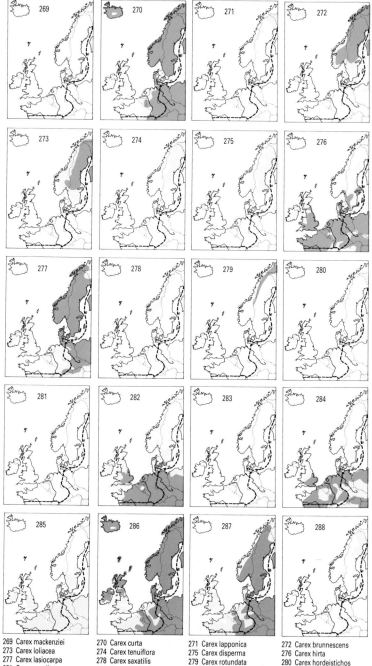

269 Carex mackenziei
273 Carex loliacea
277 Carex lasiocarpa
281 Carex secalina
285 Carex pseudocyperus

270 Carex curta
274 Carex tenuiflora
278 Carex saxatilis
282 Carex acutiformis
286 Carex rostrata

271 Carex lapponica
275 Carex disperma
279 Carex rotundata
283 Carex melanostachya
287 Carex vesicaria

272 Carex brunnescens
276 Carex hirta
280 Carex hordeistichos
284 Carex riparia
288 Carex rhynchophysa

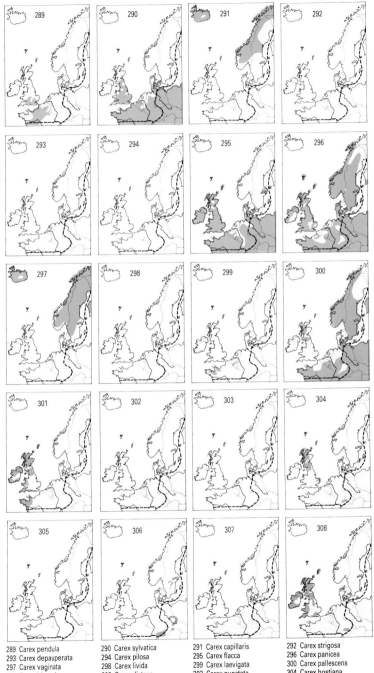

289 Carex pendula
293 Carex depauperata
297 Carex vaginata
301 Carex binervis
305 Carex extensa

290 Carex sylvatica
294 Carex pilosa
298 Carex livida
302 Carex distans
306 Carex flava

291 Carex capillaris
295 Carex flacca
299 Carex laevigata
303 Carex punctata
307 Carex lepidocarpa

292 Carex strigosa
296 Carex panicea
300 Carex pallescens
304 Carex hostiana
308 Carex demissa

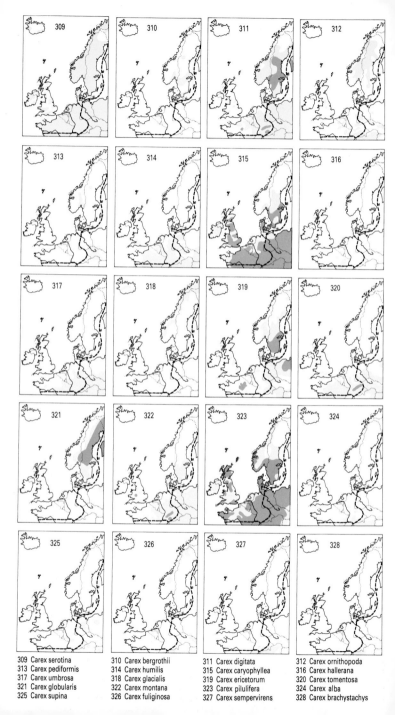

309 Carex serotina
313 Carex pediformis
317 Carex umbrosa
321 Carex globularis
325 Carex supina

310 Carex bergrothii
314 Carex humilis
318 Carex glacialis
322 Carex montana
326 Carex fuliginosa

311 Carex digitata
315 Carex caryophyllea
319 Carex ericetorum
323 Carex pilulifera
327 Carex sempervirens

312 Carex ornithopoda
316 Carex hallerana
320 Carex tomentosa
324 Carex alba
328 Carex brachystachys

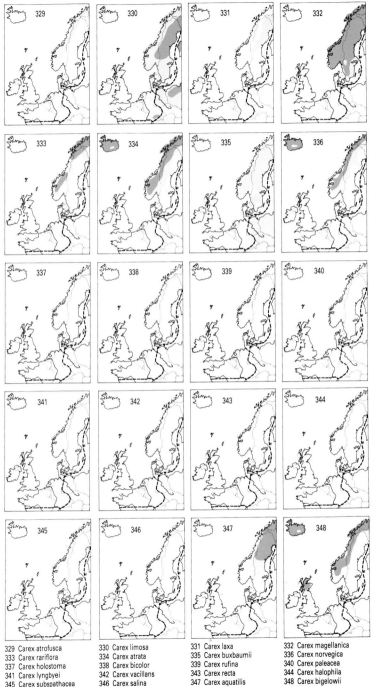

329 Carex atrofusca
333 Carex rariflora
337 Carex holostoma
341 Carex lyngbyei
345 Carex subspathacea

330 Carex limosa
334 Carex atrata
338 Carex bicolor
342 Carex vacillans
346 Carex salina

331 Carex laxa
335 Carex buxbaumii
339 Carex rufina
343 Carex recta
347 Carex aquatilis

332 Carex magellanica
336 Carex norvegica
340 Carex paleacea
344 Carex halophila
348 Carex bigelowii

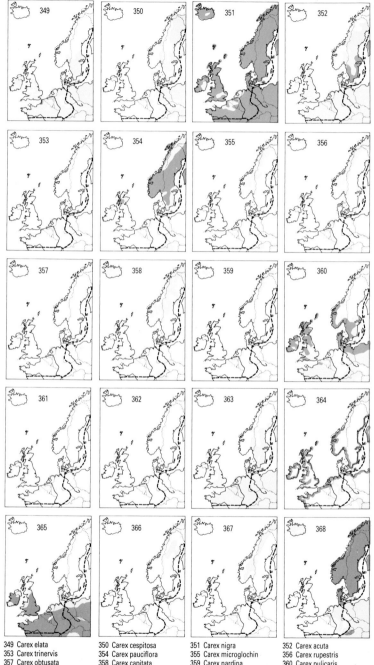

349 Carex elata
353 Carex trinervis
357 Carex obtusata
361 Juncus maritimus
365 Juncus inflexus

350 Carex cespitosa
354 Carex pauciflora
358 Carex capitata
362 Juncus acutus
366 Juncus balticus

351 Carex nigra
355 Carex microglochin
359 Carex nardina
363 Juncus compressus
367 Juncus arcticus

352 Carex acuta
356 Carex rupestris
360 Carex pulicaris
364 Juncus gerardi
368 Juncus filiformis

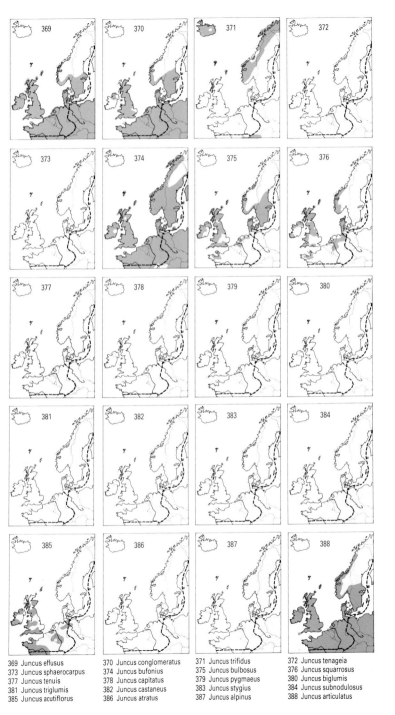

369 Juncus effusus
373 Juncus sphaerocarpus
377 Juncus tenuis
381 Juncus triglumis
385 Juncus acutiflorus

370 Juncus conglomeratus
374 Juncus bufonius
378 Juncus capitatus
382 Juncus castaneus
386 Juncus atratus

371 Juncus trifidus
375 Juncus bulbosus
379 Juncus pygmaeus
383 Juncus stygius
387 Juncus alpinus

372 Juncus tenageia
376 Juncus squarrosus
380 Juncus biglumis
384 Juncus subnodulosus
388 Juncus articulatus

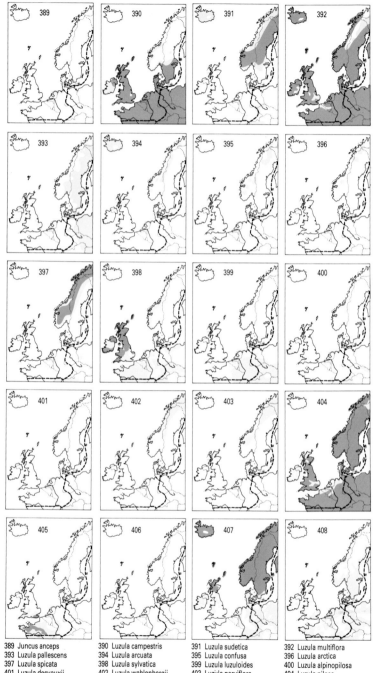

389 Juncus anceps
393 Luzula pallescens
397 Luzula spicata
401 Luzula desvauxii
405 Luzula forsteri

390 Luzula campestris
394 Luzula arcuata
398 Luzula sylvatica
402 Luzula wahlenbergii
406 Luzula luzulina

391 Luzula sudetica
395 Luzula confusa
399 Luzula luzuloides
403 Luzula parviflora
407 Huperzia selago

392 Luzula multiflora
396 Luzula arctica
400 Luzula alpinopilosa
404 Luzula pilosa
408 Lycopodiella inundata

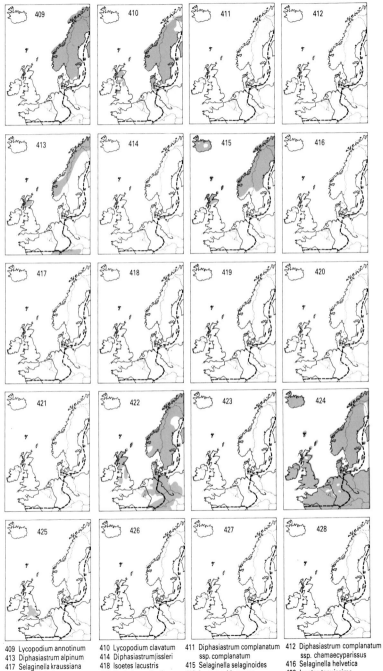

409 Lycopodium annotinum
413 Diphasiastrum alpinum
417 Selaginella kraussiana
421 Isoetes histrix
425 Equisetum telmateia

410 Lycopodium clavatum
414 Diphasiastrum issleri
418 Isoetes lacustris
422 Equisetum sylvaticum
426 Equisetum hyemale

411 Diphasiastrum complanatum ssp. complanatum
415 Selaginella selaginoides
419 Isoetes echinospora
423 Equisetum pratense
427 Equisetum ramosissimum

412 Diphasiastrum complanatum ssp. chamaecyparissus
416 Selaginella helvetica
420 Isoetes tenuissima
424 Equisetum arvense
428 Equisetum variegatum

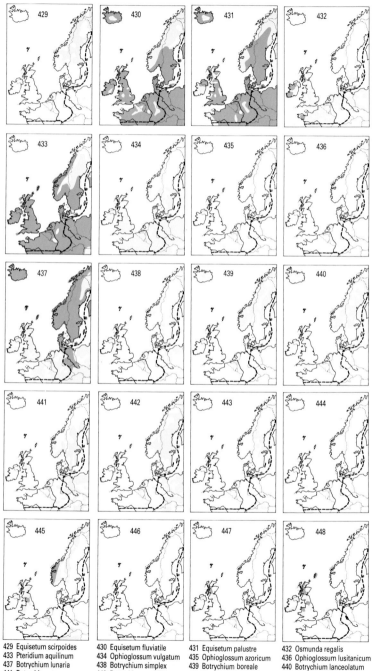

429 Equisetum scirpoides
433 Pteridium aquilinum
437 Botrychium lunaria
441 Botrychium
     matricariifolium
445 Cryptogramma crispa

430 Equisetum fluviatile
434 Ophioglossum vulgatum
438 Botrychium simplex
442 Botrychium multifidum
446 Anogramma leptophylla

431 Equisetum palustre
435 Ophioglossum azoricum
439 Botrychium boreale
443 Botrychium virginianum
447 Hymenophyllum
     tunbrigense

432 Osmunda regalis
436 Ophioglossum lusitanicum
440 Botrychium lanceolatum
444 Adiantum capillus-veneris
448 Hymenophyllum wilsonii

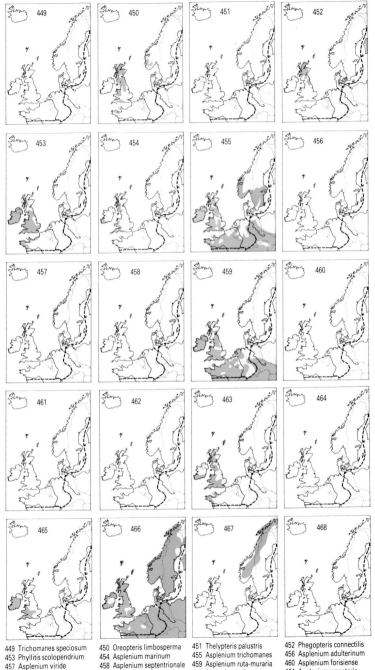

449 Trichomanes speciosum
453 Phyllitis scolopendrium
457 Asplenium viride
461 Asplenium fontanum
465 Ceterach officinarum

450 Oreopteris limbosperma
454 Asplenium marinum
458 Asplenium septentrionale
462 Asplenium billotii
466 Athyrium filix-femina

451 Thelypteris palustris
455 Asplenium trichomanes
459 Asplenium ruta-muraria
463 Asplenium adiantum-
    nigrum
467 Athyrium distentifolium

452 Phegopteris connectilis
456 Asplenium adulterinum
460 Asplenium forisiense
464 Asplenium onopteris
468 Diplazium sibiricum

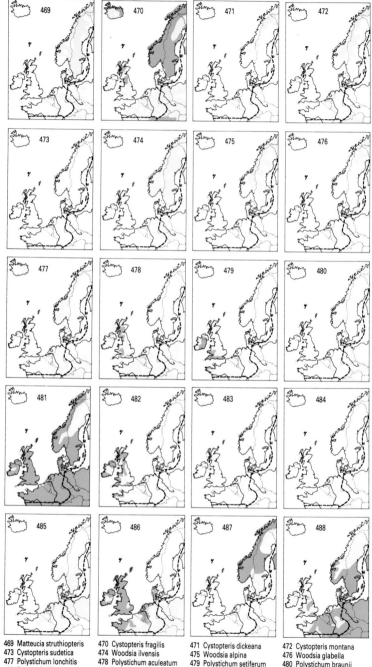

469 Matteucia struthiopteris
473 Cystopteris sudetica
477 Polystichum lonchitis
481 Dryopteris filix-mas
485 Dryopteris cristata

470 Cystopteris fragilis
474 Woodsia ilvensis
478 Polystichum aculeatum
482 Dryopteris affinis
486 Dryopteris dilatata

471 Cystopteris dickeana
475 Woodsia alpina
479 Polystichum setiferum
483 Dryopteris oreades
487 Dryopteris expansa

472 Cystopteris montana
476 Woodsia glabella
480 Polystichum braunii
484 Dryopteris submontana
488 Dryopteris carthusiana

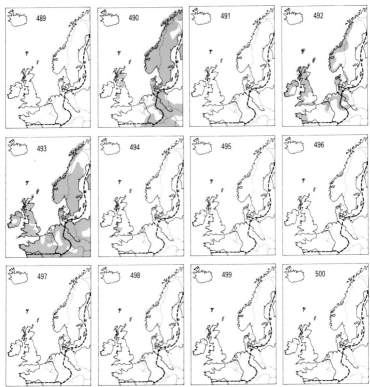

489 Dryopteris aemula
493 Polypodium vulgare
497 Pilularia globulifera

490 Gymnocarpium dryopteris
494 Polypodium interjectum
498 Salvinia natans

491 Gymnocarpium
   robertianum
495 Polypodium australe
499 Azolla filiculoides

492 Blechnum spicant
496 Marsilea quadrifolia
500 Azolla caroliniana

Further Reading

The ultimate authority for the flowers of Europe is *Flora Europaea*. The volumes relevant to this book are 1 (ferns) and 5 (grasses, sedges and rushes). There are also a number of handbooks dealing with these groups individually.

Tutin, T. G. *et al.* 1964–80: *Flora Europaea,* volumes 1–5. Cambridge: Cambridge University Press.

Grasses

Hubbard, C. E. 1968: *Grasses,* second edition. London: Penguin Books.
The most complete guide to grasses, but deals with British species only. Illustrated with black-and-white line drawings.

Sedges

Jermy, A. C., Chater, A. O. and David, R. W. 1982: *Sedges of the British Isles,* second edition. London: Botanical Society of the British Isles.
A complete guide to British sedges, with black-and-white line drawings, maps and keys, but restricted to the genus *Carex*.

Ferns

Hyde, H. A., Wade, E. A. and Harrison, S. G. 1969: *Welsh Ferns,* fifth edition. Cardiff: National Museum of Wales.
A trusty handbook; since most British ferns occur in Wales, it has proved invaluable to botanists for over 40 years.
Page, C. N. 1982: *The Ferns of Britain and Ireland.* Cambridge: Cambridge University Press.
A comprehensive new guide to British ferns.

In addition, readers are referred to the companion volume to this guide, covering the wild flowers of Britain and Northern Europe, and to the atlas of the plants in the same area.

Fitter, Richard, Fitter, A. H. and Blamey, M. 1974: *The Wild Flowers of Britain and Northern Europe,* fourth edition. London: William Collins.
Fitter, A. H. 1978: *An Atlas of the Wild Flowers of Britain and Northern Europe.* London: William Collins.

Index of English Names

252

Index of Scientific Names